Sustainability Assessment

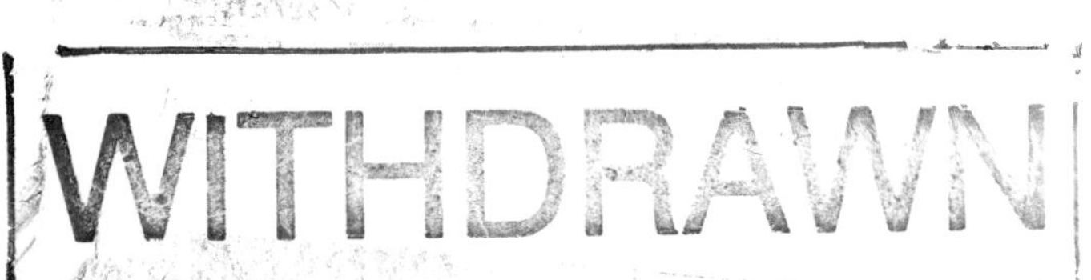

Sustainability assessment is an increasingly important tool for informing planning and development decisions across the globe. Required by law in some countries, strongly recommended in others, a comprehensive analysis of why sustainability assessment is needed and clarification of the value-laden and political nature of assessments are long overdue.

More than just case studies, this book simultaneously provides the knowledge, inspiration and range of assessment tools in decision making necessary to tackle sustainability assessment challenges nested within wide-ranging values and sustainability-grounded evidence.

This collection details the current state-of-the art in relation to sustainability assessment theory and practice, and considers the pluralistic nature of the tool and the implications for achieving sustainable decision making. The contributors set out the context for sustainability assessment and then outline some contested issues which can affect interpretations of whether the decision tool has been effective. Current practice worldwide is assessed against a consistent framework and then solutions to some of the inherent weaknesses and causes of conflict in relation to the perceived sustainability of outcomes are put forward.

The book sets out best practices in sustainability assessment through case studies and research on countries leading the way in the field. It also covers emerging factors influencing the effectiveness of decision-making tools and evaluates how they affect the performance of sustainability assessment. Written by leading academics and professionals, it is ideally suited for the growing numbers of courses in impact assessment education and training.

The Natural and Built Environment Series

Editor: Professor John Glasson

Oxford Brookes University

Real Estate: Property Markets and Sustainable Behaviour
Peter Dent, Michael Patrick and Xu Ye

Introduction to Environmental Impact Assessment 4th edition
John Glasson, Riki Therivel and Andrew Chadwick

The Environmental Impact Statement after Two Generations
Michael Greenberg

Building Competences for Spatial Planners
Anastassios Perdicoulis

Spatial Planning and Climate Change
Elizabeth Wilson and Jake Piper

Water and the City
Iain White

Urban Planning and Real Estate Development
John Ratcliffe and Michael Stubbs

Transport Policy and Planning in Great Britain
Peter Headicar

Introduction to Rural Planning
Nick Gallent, Meri Juntti, Sue Kidd and Dave Shaw

Regional Planning
John Glasson and Tim Marshall

Strategic Planning for Regional Development
Harry T. Dimitriou and Robin Thompson

Methods of Environmental Impact Assessment
Peter Morris and Riki Therivel

Public Transport
Peter White

Landscape Planning and Environmental Impact Design
Tom Turner

Controlling Development
Philip Booth

Partnership Agencies in British Urban Policy
Nicholas Bailey, Alison Barker and Kelvin MacDonald

Development Control
Keith Thomas

Expert Systems and Geographic Information Systems for Impact Assessment
Agustin Rodriguez-Bachiller and John Glasson

Sustainability Assessment

Pluralism, practice and progress

Edited by
Alan Bond, Angus Morrison-Saunders, Richard Howitt

LONDON AND NEW YORK

First published 2013
by Routledge
2 Park Square, Milton Park, Abingdon, Oxon OX14 4RN

Simultaneously published in the USA and Canada
by Routledge
711 Third Avenue, New York, NY 10017

Routledge is an imprint of the Taylor & Francis Group, an informa business

British Library Cataloguing in Publication Data
A catalogue record for this book is available from the British Library

Library of Congress Cataloging in Publication Data
Bond, A. J. (Alan James)
Sustainability assessment : pluralism, practice and progress / Alan Bond, Angus Morrison-Saunders, Richard Howitt.
p. cm.
Includes bibliographical references and index.
1. Sustainability. 2. Sustainable development. 3. Environmental impact analysis. 4. Environmental assessment. I. Morrison-Saunders, Angus, 1966– II. Howitt, Richard. III. Title.
HC79.E5.B645 2011
338.9'27—dc23
2011052239

ISBN: 978-0-415-59848-4 (hbk)
ISBN: 978-0-415-59849-1 (pbk)
ISBN: 978-0-203-11262-5 (ebk)

Typeset in Stone Serif and Akzidenz Grotesk
by Florence Production Ltd, Stoodleigh, Devon

Printed and bound by CPI Group (UK) Ltd, Croydon, CR0 4YY

Contents

List of illustrations

Boxes

Figures

Tables

Preface

Alan Bond, University of East Anglia, Angus Morrison-Saunders, Murdoch University and North-West University, Richard Howitt, Macquarie University

This book was written to empower its readers to evaluate vital aspects of the theory and practice of sustainability assessment and to contribute to strengthening future practice. It brings together a growing body of research and practice on sustainability assessment, and offers both a framework for thinking about the challenges of sustainability assessment and for evaluating specific sustainability assessment exercises.

The book originated in the collaboration between European and Australian academic institutions participating in a funded (by the European Commission and Australian Government) exchange scheme for undergraduate students in environmental studies to spend a semester overseas (see www.oceans-network.eu/). The common ground in the programs participating in that exchange was a focus on sustainability – although it might be considered something of a paradox (in sustainability terms) that the collaboration funded return flights between Europe and Australia for 95 students; that's a lot of extra carbon dioxide in the atmosphere! Of course, we hope that the benefits of the exchange experience offset the environmental impacts; writing this book certainly helps to offset the guilt of the editors who all participated in the exchange and contributed additional carbon to the atmosphere! Whether the EU–Australia collaboration has been a sustainable venture is not something we tested and, with hindsight, perhaps we should have done. This is where sustainability assessment comes in – when we need to know whether we should be doing something, or how to do something, in the best or right way.

Our work starts from the premise that sustainability assessment is inherently a good thing. We should at least try to know, *before* we do something, that it is the right thing to do. In theory, this sounds simple – but it is not. One of the reasons for this is because of the lack of clarity over our understanding of exactly what sustainability assessment is trying to achieve. For example, as of 2011, more than seven billion people inhabit the earth, and a simple internet search suggests that (depending on calculation method) around 96,000 billion humans have ever lived. So how many people will live in the future? And how many other species are there now and will be around in the future (estimates suggest there are between five and 100 million species on earth today). Out of all these people and all these species on earth, who (or what) is it that sustainability assessment is doing the right thing for? We started this book with the assumption that sustainability assessment, as currently practised, can be strengthened – and the research and practice discussed in this book will contribute to that strengthening. We think this because nowhere does current practice get to grips with this question of exactly who it is all for. Instead, each procedure has implicit definitions of what the right thing to do is and who (or what) it is for.

Furthermore, we suspect (and there is evidence to support this) that activities carried out under the name of sustainability assessment can be manipulated to provide conclusions that those in charge of the assessment want to see. If we had conducted a sustainability assessment of the undergraduate exchange scheme mentioned above, in advance, would the fact that the staff and students directly involved would benefit from free flights to somewhere they wanted to visit have biased the outcome? Probably! Even if we tried to be objective, it is likely that we would have found ways of justifying what we wanted to do by giving greater weight to the benefits accruing from the project, and lesser weight to the negative impacts. That is, the value judgements that underpin the process of sustainability assessment will inevitably affect how we judge the data and the sense that any particular assessment exercise makes of it.

In this book, we aim to consider what sustainability assessment should be doing so that we can debate what might need to change in order to achieve this. In doing so, we acknowledge the many influences on human actions as it is not easy to 'do the right thing' where it reduces the immediate and direct benefits to those conducting the assessment and/or those making the decisions. Recognition that every decision can be contested by people with valid, but contrasting, views is a key starting point for this book.

We set out with the expectation that our primary audience will be higher-education students, both undergraduate and postgraduate, with an interest in sustainability. We expect the book to be of value to other audiences too, particularly sustainability assessment practitioners, government regulators, researchers and anyone who might be affected by planning activities in any sector. As the primary audience is students, we have written the book accordingly and emphasise the importance of research skills: in particular, the ability to see the other perspectives and not take things for granted. We aim to be critical in order to move sustainability assessment theory and practice forward. However, we would urge the reader to distinguish between critical appraisal (which is what we attempt to provide) and criticism. The former is the backbone of any research and subsequent policy making; the latter is often a biased value-judgement with no benefit for policy making. In considering sustainability assessment in this book, we have been very critical, but this should not be considered to be negative. On the contrary, we are very positive about the fact that sustainability assessment is practised at all and recognise that it is still in the early stages of its evolution. It will get better, but it is only through critical reflection that improvement can occur.

In appraising practise, we have endeavoured to adopt the critical stance of a person who considers a glass to be half full rather than half empty. That is, we appreciate the advances that have been made so far in this fledging endeavour and, in highlighting limitations (as you should), acknowledge these in light of the pressures placed on any researcher or practitioner in terms of time and resources, or political pressure. We also acknowledge that the sustainability assessment processes adopted in various countries have often been accepted in good faith by practitioners, potentially unaware of inherent flaws. Thus, we have designed this book to advance sustainability assessment and hope that it will make a valuable contribution to the improvement of this decision-making tool. We hope you are up for the challenge to advance practice accordingly and make your own direct contribution to a more sustainable world!

The structure of the book

We have divided the book into four separate parts, each with their own chapters. The reasons for dividing the book into these parts, and the content of the chapters within them, are detailed below.

Part 1 Introduction

The first part, the introduction, sets the context for the rest of the book in examining the rationale for conducting sustainability assessment in the first place (because, otherwise, there would be no need for the book). It also investigates the theory of environmental assessment, from which sustainability assessment has developed. The theory and historical background introduced here is important to understand as it underpins practice and is essential to the discussion that follows. If supporting sustainable development is the goal for sustainability assessment, practitioners need to understand the many theories about how best to go about this, and the reasons for and implications of different theoretical approaches. Whichever theory (or theories) is adopted will affect the outcome of both the assessment and implementation phases of any particular project or action. For example, it might be theorised that sustainability assessment needs to consider alternatives to the proposed actions in order to properly predict and evaluate the likely effects – and so this would be a required procedural stage. So theory determines what sustainability assessment actually does.

Chapter 1, by Bob Gibson, answers the question *why sustainability assessment?* It examines, by way of historical context, the roots of impact assessment born out of the environmental movement and first legislated in the United States in 1969. It traces the roots of arguments for sustainable decision making and examines, in particular, the current status of decision making with respect to sustainable outcomes. Bob argues that the world is engaged in largely unsustainable practices that demand some action, and pretty soon.

Chapter 2, by Mat Cashmore and Lone Kørnøv, consider the *changing theory of impact assessment* and argue that impact assessment is a singularly theoretically underdeveloped field of endeavour. This chapter examines how our theoretical understanding has developed over time, moving from a period when it was assumed that all decision makers acted as rational individuals with decisions based solely on the scientific evidence, through to a more contemporary emphasis on the legitimacy of plural interpretations of its objectives and acceptable ways of achieving them. The implications of different theoretical interpretations are explained.

Part 2 Sustainability assessment: pluralism

The second part of the book focuses on sustainability assessment pluralism. By this, we mean the different possible interpretations of a number of key issues relating to the outcomes of sustainability assessment.

Chapter 3, by Alan Bond and Angus Morrison-Saunders, examines *challenges in determining the effectiveness of sustainability assessment.* This sets the scene for this part of the book by detailing a number of contested definitions which complicate the interpretation of sustainability assessment. In particular, the two main paradigms for sustainability – weak sustainability versus strong sustainability – are considered. Methodological paradigms are introduced, including the issue of reductionism (reducing the sustainability implications to discreet, manageable parts) versus holism (considering sustainability implications in the whole); opening up (providing decision makers and other stakeholders with the complete, often conflicting picture) versus closing down (providing a single solution in terms of what should be done). The pluralistic nature of the meaning of effectiveness, in relation to sustainability assessment, is also covered.

Chapter 4, by Rich Stoffle, Brent Stoffle and Annelie Sjölander-Lindqvist, analyse *contested time horizons* through drawing on the uncertainty that is inherent in the definition of intergenerational equity (and intragenerational equity). They examine how different stakeholders view time differently and consider its cultural importance to some groups within society. The critical issue in this chapter is the challenge for sustainability assessment given time horizons favoured by decision and policy makers as opposed to other stakeholders and citizens.

Chapter 5, by Richard Howitt, examines *contested spatiality: geographical scale in sustainability assessment.* Any development, plan or policy has implications at different spatial scales: usually the scales over which effects are experienced are different for different sustainability issues. The chapter examines the implications of these variations for data handling and for the application of a sustainability assessment framework.

Chapter 6, by Donna Craig and Michael Jeffery, examines *legal pluralism – notions of standing and legal process constraining assessment.* Decisions, and decision-making processes, are ultimately controlled by courts and it is, therefore, crucial that the role courts play in interpreting sustainability assessment is examined. The chapter focuses on particular legal contexts and explores the critical issue of legal standing which stipulates who has, and does not have, direct access to courts in relation to the sustainability assessment process and associated decision-making process.

Chapter 7, by Jenny Pope and Angus Morrison-Saunders, examines *pluralism in practice.* This chapter, written from the standpoint of a practitioner, acknowledges the different positions that can be adopted by different stakeholders in sustainability assessment, which should lead to recommendations for greater deliberation. But what happens when the 'chips are down' and how do those stakeholders directly involved in the sustainability assessment deal with pluralism in practice?

Part 3 Sustainability assessment: practice

The third major part of the book critically appraises what happens now in terms of sustainability assessment practice. Early examples of sustainability assessment can be found in England, Western Australia, Canada and South Africa. These examples are very different in terms of political and geographical context, and what they are applied to.

Chapter 8, written by Alan Bond, Angus Morrison-Saunders and Richard Howitt, sets out a *framework for comparing and evaluating sustainability assessment practice* against which the content of the subsequent practice examples will be compared. The purpose of this chapter is to set a common framework enabling cross comparison, and to ensure that elements of sustainability assessment which might be considered critical are appraised in the rest of the chapter.

Chapter 9, written by Riki Thérivel, examines *sustainability assessment in England.* The practice of sustainability assessment being conducted for spatial plans at a number of geographical scales has developed over a number of years, and it has been possible to examine the effect of sustainability assessment on the development of these plans.

Chapter 10, written by Angus Morrison-Saunders and Jenny Pope, focuses on *learning by doing: sustainability assessment in Western Australia.* With a strong tradition of project-based approaches to impact assessment driven by the resource sector, this chapter will showcase the strengths and weaknesses of sustainability assessment arising from this foundation. Several case studies from a range of sustainability assessment applications in recent practice are presented.

Chapter 11, written by Bob Gibson, examines *sustainability assessment in Canada.* The chapter traces the history of sustainability assessment in Canada, from precursors to processes which could be considered to be assessing sustainability implications, through to the present day, including the overall character, resistance and prospects for sustainability assessment at the strategic and project levels.

Chapter 12, written by Francois Retief, examines *sustainability assessment in South Africa*. The consideration of a jurisdiction which has legislated the concept of sustainability widely, but has not formalised the need for any specific process of sustainability assessment, is particularly interesting because there are fewer constraints on the problems of pluralism than in other countries with established procedures.

Part 4 Solutions?

Based on the critical appraisal of current practice, and in light of the pluralism inherent in many aspects of sustainability assessment, the final part of the book proposes new directions for sustainability assessment through a focus on improving some of the key weaknesses.

Chapter 13, written by Ciaran O'Faircheallagh and Richard Howitt, focuses on *better engagement*. By definition, pluralism suggests that different views or values are held by different stakeholders. In order to accommodate all of these, it is essential that engagement is fully inclusive and effective, and the authors draw on their experience in the field of social impact assessment to indicate how to engage effectively.

Chapter 14, written by Alan Bond and Angus Morrison-Saunders, addresses the concern that a good process should not simply lead to sustainable outcomes in the isolated case in question. Ideally, it should also lead to *better learning* whereby individuals and institutions learn from their involvement in the process and, perhaps, change their attitudes and/or values as a result. This can create significant potential for sustainable outcomes to pervade the goals of individuals and institutions in all activities, not necessarily just those associated with sustainability assessment.

Chapter 15, by Alan Bond, Angus Morrison-Saunders and Gernot Stoeglehner, explores *designing an effective sustainability assessment process*, and how this might be achieved. This chapter examines the strengths and pitfalls of sustainability assessment processes and relates these to sustainability outcomes. It addresses many of the issues identified throughout the book, and identifies the connections between the various types of effectiveness as a means for suggesting principles which can lead to better sustainability assessment process.

Chapter 16, by Gernot Stoeglehner and Georg Neugebauer, focuses on *integrating sustainability assessment into planning: benefits and challenges*, and aims at putting sustainability assessment in its decision-making context. The chapter addresses the question of what can be expected from sustainability assessment for plans and programmes, taking into account the integratedness (integration of environmental, social, economic aspects) of plans or programmes (PP).

The strategy to achieve sustainable planning might not be to introduce a sustainability assessment scheme but to work on planning schemes that promote 'integrated' planning in line with sustainable development. This is discussed from a planning theory perspective.

Chapter 17, by Alan Bond, Angus Morrison-Saunders and Richard Howitt, presents our conclusions in terms of *applying learning to subsequent practice*. This chapter presents the key lessons learnt from sustainability assessment theory and practice arising throughout the book and provides sustainability assessment students and practitioners with practical guidance on the way forward.

The book structure aims to build a sequential argument and therefore it is intended that the chapters are read in order. However, we recognise that some readers will dip in as appropriate to their own interests or needs. Our only caution, which also applies to those who read the book in order, is that pluralism affects practice and progress. Everyone has their own view of sustainability. The authors of chapters of this book and our readers will not necessarily agree with or take the same views, and Chapter 3 emphasises this point and provides a framework for ensuring different views are considered.

Our key message is that pluralism is not a barrier to good sustainability assessment: it is a fact of life. As such, it is something which needs to be acknowledged and integrated into practice in order to progress. This book highlights pluralism with the hope that embracing its existence, and accepting and accommodating the validity of different perspectives, will advance sustainability assessment processes, practice and performance.

About the editors

Alan Bond has more than 20 years' experience in impact assessment. He is a member of the Editorial Board of *Environmental Impact Assessment Review* and *Impact Assessment and Project Appraisal*, and is an Institute of Environmental Management and Assessment Quality Mark Review Panel member. Alan sits on the Radioactive Waste Management Directorate's (part of the Nuclear Decommissioning Authority) Environmental Assessment Advisory Panel and has previously conducted research for the WHO, European Commission, Health Development Agency, European Environment Agency, Welsh Assembly Government, and Environment Agency, and has undertaken work with a number of consultancies on impact assessment. More recently, he has published widely on sustainability assessment and the challenges of integrating social, economic and environmental considerations into decision making.

Angus Morrison-Saunders is Senior Lecturer in Environmental Assessment, Murdoch University, Australia and Extraordinary Professor in Environmental Sciences and Management, North West University, South Africa. He is also co-Director of Integral Sustainability, Western Australia and co-Editor of *Impact Assessment and Project Appraisal* (the journal of the International Association for Impact Assessment). Angus has over 20 years' experience in impact assessment. Having originally been interested in evaluating the effectiveness of environmental impact assessment, in recent years he has focused on sustainability assessment theory and practice and he teaches, runs training courses, researches and conducts consultancies in this rapidly emerging field.

Richard Howitt is Professor of Human Geography at Macquarie University, Sydney. He teaches in the planning, environmental and geography programs, leads Macquarie's postgraduate SIA program, and is director of the university's research relationship with the city of Ryde, the university's host local government authority. His research deals with the social impacts of mining on Indigenous peoples and local communities, and is generally concerned with the interplay across scales of social and environmental justice, particularly in relation to indigenous rights. He received the Australian Award for University Teaching (Social Science) in 1999 and became Fellow of the Institute of Australian Geographers in 2004.

Part 1

Introduction

1 Why sustainability assessment?

Robert B. Gibson, University of Waterloo

> The scientific evidence tells us all we need to know: carry on with business-as-usual growth-at-all-costs, and we're stuffed.
>
> Jonathon Porritt (2009)

> The future ain't what it used to be.
>
> Yogi Berra (Berra and Kaplan, 2002)

Introduction

The answer to the title question may seem obvious. We should have sustainability assessment because, with a few salutary exceptions, what we are doing on Earth is wrecking the place. The dominant trends of human activities and their effects are towards deeper unsustainability. Probably for our own foreseeable interests, certainly for those of our children and undoubtedly also for many other life forms on the planet (excepting cockroaches and their numerous ilk), we need to reverse direction.

That is both why and to what end we should be adopting and applying sustainability assessment broadly, consistently and devotedly. The core rationale and fundamental standard for sustainability assessment should be to ensure that every one of our potentially significant undertakings is designed to deliver positive contributions to sustainability – multiple, mutually reinforcing, fairly distributed and lasting positive contributions to sustainability – while avoiding persistent damages.

Sadly, we are doing very little of the sort. And not much of what is now labelled sustainability assessment aspires far in that direction. Mostly, what we have so far are promising signs and untapped potential.

Over the past couple of decades, many sustainability assessments have been undertaken, under a wide variety of names, by many different bodies in an expanding host of jurisdictions. A rapidly expanding diversity of national, regional and local governments and government agencies has adopted assessment and planning processes that are in some way sustainability oriented. So have countless corporations, universities, product labelling organizations, multilateral banks and other institutions. The quality of the results has not yet been subject to rigorous evaluation. This book may be the first to attempt a general 'state of the art' appraisal and it cannot possibly cover all of the many versions of sustainability and assessment. But even the most cursory review reveals that many interests have put the sustainability label on assessment initiatives that are too narrow in scope, too selective in areas of attention, too poorly integrated and too focused on the near term to merit the title. It will also be apparent in the following chapters that even the most advanced official sustainability assessment regimes have serious limitations.

So far as I know, not one established assessment law, formal process or corporate practice approaches the basic, simple and logical standard for reversing unsustainable trends sketched above. Not in theory, much less practice.

Instead, most of our shining examples – assessments that embody serious efforts to reverse unsustainable trends and deliver multiple, mutually reinforcing, fairly distributed and lasting positive contributions to sustainability – are the product of *ad hoc* initiatives. Many are small scale. Typically they are the works of exemplary individuals, organizations and communities – nuns establishing a largely self-sufficient school for poor and disabled children in rural Tanzania (John, 2009), an alternative development organization working with the people and residual resources of a cluster of villages in India (Vaidyanathan, 2002), a green group delivering low-income residential energy efficiency in Montréal (Ribaux, 2000). We can all celebrate our favourite cases and hope that their tribes increase. But they remain exceptional, unique and disconnected. The assessment element is usually so deeply buried in an iterative process of deliberation, experiment and successive decisions that it cannot be distinguished as a particular activity. Though they may address global challenges, the best examples typically focus on the specific unsustainabilities of a place. They reflect unique combinations of local imperatives, recognized interconnections and engaged players. They are mostly driven by civil society, rather than by governments or the private sector. And they are well out of the mainstream.

That is not necessarily reason for despair. Small and local is generally more nimble than big and ponderous. Huge, intertwined, path dependent and deeply entrenched institutionalized systems are not likely to respond positively and quickly to highly inconvenient realities that are largely of their own making. In the circumstances, it is remarkable that so many major institutions have adopted the language of sustainability and have initiated sustainability assessments in some form, however imperfect those forms may still be. We can also find grounds for optimism in the great diversity of inspiring small, local, outside-the-mainstream applications. They are demonstrations of what is possible and they are valuable reminders that success in sustainability assessment, as in so much else, depends heavily on deep understanding of the peculiarities and possibilities of particular places.

The 'why' of sustainability assessment is therefore deeply entwined with the 'how' and the 'where'. The basic, simple and logical overall purpose of sustainability assessment may rest on a set of evident global imperatives and a broad understanding of what kinds of responses are required. But any potentially successful pursuit of this overall purpose depends on two dimensions of integration in practice. The first is integration of the usually separated overall concerns about human and ecological wellbeing (conventionally presented as the economic, social and biophysical pillars of sustainability). The second is integration of the global imperatives and understandings with the more specific purposes, concerns, pressures and opportunities recognized by particular organizations and people in particular places.

We can see the gradual beginnings of such integrations in the emergence of sustainability assessment and its advances so far. Arguably there is a discernable trajectory of this sort in the usual timeline landmarks in the evolution of environmental assessment and the sustainability agenda (see Environmental assessment and sustainability timeline, pp.16–17). But these landmark steps are not indicators of an automatic process. All progress to date has been a struggle against resilient institutional barriers – including habits of thought as well as structures of power and authority – and future progress is likely to depend on understanding both what is needed and what needs to be overcome.

This introduction therefore begins with the problem of deepening unsustainability mostly for the purposes of context. The discussion will be devoted mostly to exploring the main implications for movement towards sustainability-based decision making, understanding the inevitable resistance and identifying the key characteristics of sustainability assessment that need to be embraced in the mainstream before all light in the tunnel disappears.

Trends and implications

For most people on almost every continent, life has been getting better, at least materially. Over the past couple of hundred years, in many places over the last few decades, working people have gone from mostly subsisting to mostly comfortable. Life expectancy at birth is well up. So, overall, are literacy, access to birth control, democracy (very loosely defined) and tolerance of difference. In recent years, our current institutions have delivered technological advances in resource efficiencies and pollution control, better communications, broader engagement in governance and successes against many of the old communicable diseases. They have moved fingers further away from the nuclear button. They even deserve some credit for the initial steps taken so far towards effective sustainability assessment. All these and more are to the good.

The catch is in the consequences. When in 1798 the gloomy and uncharitable curate Thomas Robert Malthus predicted rising British population would inevitably crash against limited food resources (Malthus, 1970), he missed a few key considerations. He failed to anticipate advances in technology and industrial efficiencies. He did not imagine that increased wellbeing might lead to lower birth rates, at least if the poor got some share and women received some rights (both of which Malthus also failed to imagine). And he did not foresee a world of global trade exploiting the whole planet's resources.

We can now factor those considerations into the analysis, along with some other possibilities and worries that Malthus missed. The results are not encouraging. Several major prevailing trends point to deepening unsustainability. Three are key:

(i) Demands on biospheric carrying capacity are rising. Growing demands for energy, food and other commodities have been leading to more greenhouse gas emissions, land use changes, water cycle disruptions, loadings of synthetic and other contaminants – all of them pressures on depletable resources and on the assimilative and resilience capacities of ecosystems. Already visible effects include the continuing decline of indigenous biodiversity (WWF, 2010; Butchart *et al.*, 2010); changes in atmospheric chemistry and climate behaviour (IPCC, 2007); depletion of crucial resources including groundwater, ocean fisheries, soil quality and conventional oil; and adoption of increasingly costly and/or risky technologies (e.g. nuclear power plants in active seismic zones).

(ii) Most benefits of economic growth flow to those already advantaged. Despite overall GDP growth, global and national inequities in basic measures of wellbeing remain deep (UNDP, 2010). The richest 10 per cent of the world's population get about 67 per cent of the world's income while the poorest 10 per cent get about 0.22 per cent (Milanovic, 2005). Meanwhile, about 925 million people were malnourished in 2008, up from about 850 million in 1990 (FAO, 2008: 2, 48); 2.7 billion people live on less than $2/day (World Bank, 2004) and prospects of meeting the eight Millennium Development Goals for ending poverty by 2015 are declining (UN, 2010).

(iii) The likelihood of deepening tensions, conflicts and cascading system failures is growing. Some of these result from the now globally as well as regionally interactive effects of biospheric stresses and grievances in an inequitable world. But these also combine with difficulties due to other vulnerabilities and uncertainties, including the fragility of important aspects of the international financial and food systems, broad accessibility to significant means of destruction, and perceptions of insecurity and vulnerability that encourage narrowly defensive preferences.

These trends pose big challenges for global futures and, by unavoidable consequence, for regional and local futures. They also pose big challenges for current decision makers. The basic message from the three big trends is that we are now clearly in a world of practical limits and complex interdependencies – two realities that rattle the fundamental assumptions that have been informing our most powerful decision-making institutions.

Conventional decision-making institutions have assumed that we can improve overall wellbeing through economic growth based on further increases in the exploitation of energy and material resources; that the trickling-down of expanding material wealth will gradually eliminate poverty; that economic motives and technological innovation can be counted upon to find substitutes for everything we degrade or deplete; and that we can rely on impact mitigation to protect valued ecological and socio-cultural qualities. On a planet where we have apparently already overshot the sustainable carrying capacity for humans (given our current technological and managerial abilities), these convenient old assumptions are no longer valid.

Similarly, on a planet where the threats to human wellbeing, ecological integrity and climate stability are deeply entwined, highly complex and globally significant, it is not safe to assume that narrowly specialized experts will have the necessary understanding, that self-interested nations and commercially focused corporations will grasp the broader public interest, or that any set of simple answers will apply adequately to the great diversity of needed applications. And on a planet where the most powerful public- and private-sector bodies have not been able to arrest our slide into ever deeper unsustainability (much less ensure ecological recovery, eliminate poverty and stabilize atmospheric chemistry), it is no longer sensible to depend on them, at least not as presently mandated and motivated. Here too, new thinking and new practices are needed.

That is not to say that all the old ideas are to be pitched in the dustbin of history. After all, they did deliver important gains, and they continue to play important roles. But if we are to reverse direction and move towards sustainability, the old ways need to be adjusted, supplemented and subsumed by some crucial new approaches that recognize the ultimately limited capacities of our planetary resources and biophysical systems, cut back on our overall material and energy demands, and provide enough for all. The basic requirements for transition to more desirable and durable future possibilities are set out a little more rigorously in Box 1.1.

For sustainability assessment purposes, these eight requirements can be translated into generic criteria for determining which options for new or renewed undertakings promise the greatest overall contribution to sustainability. The generic criteria are not enough. They need to be elaborated and specified for application to particular cases and contexts (Gibson 2006, 2011), and they need to be supplemented by similarly specified guidance for assessing trade-offs (Gibson *et al.*, 2005: ch.6). But they provide a useful indication of the profound difference between assessments that are seriously committed to applying a 'contribution to sustainability' test, and the more conventional assessments that aim only to mitigate significant adverse effects. Serious sustainability assessment is not about making our activities somewhat less damaging. It is about reversing the unsustainable trends. Consequently it is not about minor adjustments. It is about transition.

Box 1.1 Eight requirements for progress towards sustainability

Socio-ecological system integrity

Build human–ecological relations that establish and maintain the long-term integrity of socio-biophysical systems and protect the irreplaceable life support functions upon which human as well as ecological wellbeing depends.

Livelihood sufficiency and opportunity

Ensure that everyone and every community has enough for a decent life and opportunities to seek improvements in ways that do not compromise future generations' possibilities for sufficiency and opportunity.

Intragenerational equity

Ensure that sufficiency and effective choices for all are pursued in ways that reduce dangerous gaps in sufficiency and opportunity (and health, security, social recognition, political influence, etc.) between the rich and the poor.

Intergenerational equity

Favour present options and actions that are most likely to preserve or enhance the opportunities and capabilities of future generations to live sustainably.

Resource maintenance and efficiency

Provide a larger base for ensuring sustainable livelihoods for all while reducing threats to the long-term integrity of socio-ecological systems by reducing extractive damage, avoiding waste and cutting overall material and energy use per unit of benefit.

Socio-ecological civility and democratic governance

Build the capacity, motivation and habitual inclination of individuals, communities and other collective decision-making bodies to apply sustainability principles through more open and better informed deliberations, greater attention to fostering reciprocal awareness and collective responsibility, and more integrated use of administrative, market, customary, collective and personal decision-making practices.

Precaution and adaptation

Respect uncertainty, avoid even poorly understood risks of serious or irreversible damage to the foundations for sustainability, plan to learn, design for surprise and manage for adaptation.

Immediate and long-term integration

Attempt to meet all requirements for sustainability together as a set of interdependent parts, seeking mutually supportive benefits.

(Gibson *et al.*, 2005: ch.5)

Transitions and institutions

The trends towards deeper unsustainability are products of the old assumptions, the prevailing ways of defining problems and solutions, and the established means of organizing power and authority. These institutionalized structures and practices are intricately intertwined, mutually reinforcing and firmly entrenched. While they have brought significant improvements in wellbeing for many people in many places, the most powerful established institutions are mostly designed and committed to maintaining the drivers of conventional economic growth and conventional distribution of benefits. They are not equipped, or as a priority inclined, to respect biospheric limits or to deliver sufficiency for all.

The pursuit of sustainability therefore involves fostering fundamental transitions in our authoritative institutions as well as transitions in the particular practices that are driving undesirable trends in planetary economy and ecology. Fostering institutional transition is a core role of sustainability assessment. Its means of accomplishing this is by requiring and guiding adoption of a way of thinking, planning, evaluating, choosing and acting that is fundamentally different from conventional approaches – so different that it will result in the undertakings that contribute to sustainability rather than to the prevailing trends.

Moreover, the necessary transitions are for the planet as a whole. The trends towards deeper unsustainability are global. So is the reach of many of the institutions and undertakings involved. Humans have caused many smaller scale socio-ecological collapses over the years, but in the past there were always other places and populations for recovery (Diamond, 2005). Failure to live sustainably at the global scale has deeper implications. We have no back-up planet. The practical imperative, then, is to foster transitions in institutional thinking and behaviour everywhere.

At the global level we need shared understanding and commitment to meeting the broad requirements for progress towards sustainability – some basic common appreciation of the global challenges and what we have learned so far about responding to them. At the local/regional/national levels we need specific processes to apply sustainability thinking in decision making on a host of different matters, including the strategic and project initiatives that are normally identified as proper subjects of assessment requirements. We also need to recognize the close-to-home threats to longer-term wellbeing, and the lessons from attempts in our own backyards to find positive and lasting solutions.

The resulting agenda for sustainability assessment is highly ambitious because that is what the situation demands. It is also, unavoidably, daunting and difficult. Progress will be slow. No serious transition comes easily and quickly. Few governments or other organizations with entrenched ideas and practices will happily require themselves to think and act differently. They will need to be encouraged, pushed, perhaps even tricked. Opportunities for innovation will have to be found in places where the most hidebound authorities are not paying attention, in circumstances where the old approaches have clearly failed, in cases where the issues are too new to have been captured under defined mandates, or where new stakeholders have elbowed their way to the table.

Possibilities are likely to be greater where damages and hazards are more immediate and more obvious; where responsibility can be pinned on identifiable and at least somewhat accountable authorities; and where the problems can be treated as opportunities for positive change.

The greatest advantage of the sustainability assessment agenda is that it is positive. Everyone wants a more desirable and durable future. That goal is obviously more attractive than mitigating the most adverse effects. It may be unconventional. It certainly involves more work – considering more factors, listening to more voices, looking further ahead. But if all that can be done with reasonable fairness and efficiency, and if it can deliver actual gains in the long run, it will be embraced.

The evident lessons

Nothing important is simple. Indeed, nothing important is merely complicated. Our world and all of our big challenges in living in it are complex, minimally predictable and richly intertwined.

Brenda Zimmerman has done a nice job of clarifying the differences: following a recipe to bake a cake, she says, is simple; building a rocket to get a human to the moon and back is complicated; raising a child is complex (Glouberman and Zimmerman, 2002). Helping materially deficient communities and nations to achieve reasonable wellbeing is complex. So is developing suitable responses to climate change, avoiding pandemic risks, reducing food system vulnerabilities and establishing a resilient and equitable global financial system.

The unsustainable global trends that we must now address are, in part, the product of simplifying assumptions that crashed into the actual complexity of the real world. For administrative purposes life would be simpler if growth automatically delivered development and if living on a single planet imposed no limitations. Life would be complicated but manageable if we could deal with our problems effectively by breaking them down into digestible bits and particular responsibilities and imagining the interconnections were trivial.

In the actual world, we need to deliver growth in material wellbeing for many people, while also reducing our demands on the planet's carrying capacity. We need to enhance ecological stewardship in ways that also deliver viable livelihoods. We need to multiply efficiencies in the use of energy and other resources but also retain flexibility, diversity and redundancy to minimize system vulnerabilities. We need to foster innovation but also minimize exposure to poorly understood risks of ambitious interventions. We need to initiate significant collective initiatives for transition to more sustainable paths but use these to establish cultural changes that markets and governments cannot possibly accomplish on their own. We need to act within the small part of the world that we can claim to understand, but do so in ways that address global challenges in common cause with communities we will never visit and generations not yet born.

Sustainability assessment can be useful only insofar as it responds to these actual needs of the real world. It must aim to reverse our drift towards the planetary precipice. It must foster a transformation, rather than merely mitigate the continuing damage. That is the foundation of sustainability assessment's basic, logical role in ensuring that every potentially significant undertaking is designed to deliver lasting positive contributions to sustainability – multiple, mutually reinforcing and fairly distributed – while avoiding persistent damages. But this is just the beginning of the required subversion.

Sustainability assessment must also embrace complexity, respect context, favour broad engagement, recognize interconnections and care for the future in a world where in all cases the established institutional biases prefer the opposite.

The good news is that sustainability assessment is just one area of progressive initiative among countless others that are facing the same realities and recognizing the same imperatives. Consider how the conservation movement has learned that desired species cannot be protected without maintenance of ecosystem habitats and viable livelihoods for the local people. Or how urban planning has evolved from accommodating expansion to designing alternatives to sprawl. Or how civil society organizations have by-passed governments to establish certification systems for more responsible processes and products. Or how demand management has entered the mainstream in energy and water policy. Or how economists are now attempting to value ecological goods and services. Or how best practice in conventional environmental assessment has come to incorporate attention to uncertainty and precaution, cumulative and legacy effects, and application to strategic undertakings.

Even the peculiar history of 'sustainable development' indicates that the bigger agenda is gradually elbowing its way into practice. The notion, popularized by the World Commission on Environment and Development's 1987 report (WCED, 1987), was widely embraced as a fuzzily positive alternative to accepting limits to growth. But it planted the idea that poverty and ecological degradation needed to be confronted together and that a host of other powerful interdependencies were involved. Gradually, over the following decades, all of the vague government and corporate commitments have raised expectations of improved behaviour. The initial symbolic statements are now beginning to be incorporated in enforceable law. Private sector firms that have been pushed into sustainability reporting and claims to corporate social responsibility are now finding (often to their horror) that these are evolving into specified obligations.

Most of the progress is tentative. None of it is comprehensive or firmly established. Ecological considerations are still mostly marginal. There are few efforts to ensure that the benefits of growth flow mostly to the least advantaged. But there are certainly indications of a positive co-evolution of theory and experience in a wide variety of fields. The common characteristics suggest acceptance of needs to address a complex, diverse and increasingly stressed world and willingness to replace narrow and reductionist solutions with approaches that involve a wider range of options, participants and objectives.

Together, these considerations provide the bigger and more powerful but also messier answer to 'why sustainability assessment?' We are stumbling towards more informed, capable and ambitious sustainability assessment because of a host of interacting factors that have emerged from the basic failures and lessons of the past century. And we are working on more farsighted, broadly integrative and context-sensitive ways of planning new undertakings for many reasons. At the base, however, is a recognition that deliberations on new projects, policies, plans and programmes are important venues for a positive transition based on the co-evolution of ideas and experiments with ways of living in a world that is complex, surprising, diverse and vulnerable.

Six imperatives for sustainability assessment

We need sustainability assessment because the main prevailing human-induced trends of our time are suicidal as well as damaging to many other living things – we are expanding our demands on the biosphere even though they are already beyond what the planet can maintain, and we are delivering most of the benefits of this expansion to those who already have at least enough, when maybe a third of the human population doesn't have enough. If these trends are not reversed, things will get increasingly and perhaps irretrievably ugly.

But that is not why we have the beginnings of sustainability assessment practice today. Sustainability assessment is emerging as one of many interrelated responses to the lessons of the past. Most of these are lessons from the failures of conveniently simple ideas in an inconveniently complex world. Like the other responses, sustainability assessment recognizes that there are no easy answers. There is no way of avoiding the intricate and powerful connections among social, economic and ecological factors or local and global scales, or short- and long-term implications. There are no general theories that can safely ignore the particulars of the case and place. There is no automatic mechanism that will deliver a better future. And there are few grounds for lasting confidence in a world that we understand so poorly.

Combined, the need to reverse prevailing trends and the lessons from past simple-mindedness and hubris point not only to the main reasons for sustainability assessment but also to how sustainability assessment should be conceived and what it should aim to deliver. We can end this introduction and begin the book with six imperatives concerning the why, what and how of sustainability assessment:

(i) Sustainability assessment must aim to reverse the prevailing trends towards deeper unsustainability by insisting that every one of our new or renewed projects, programmes, plans and policies makes a positive contribution to a desirable and durable future. The usual environmental assessment objective is merely to mitigate significant adverse effects. There is no long-term hope in that. Mitigation can only slow our slide over the precipice when what we need is to reverse direction.
(ii) Sustainability assessment must ensure integrated attention to all of the key intertwined factors that affect our prospects for a desirable and durable future. That will still entail particular studies of usually neglected individual factors – ecological integrity and equity effects, for example. But in a world where interactive effects are typically crucial and where most powerful institutions have little capacity and less inclination to work across the boundaries of mandate and expertise, effective attention to interactions and informed integration of considerations will be accomplished in sustainability assessment or not at all. Moreover, the greatest positive prospects lie in the linkages.
(iii) Sustainability assessment must seek mutually reinforcing gains. What we face are vicious cycles of ecological degradation and resource depletion, which undermine livelihoods, which increases desperation and conflict, which further undermines cooperation and foresight and stewardship, which leads to further degradation and depletion, and so on downward. In response, sustainability assessment must be a vehicle for seeing the interdependence of ecology, economy and society and for finding ways to serve all three at once, in ways that are mutually reinforcing so that we can generate virtuous circles, spiralling upward.
(iv) Sustainability assessment must seek to minimize trade-offs. It is not about balancing ecology, economy and society as competing priorities. That can deliver only compromises, with the usual sacrifices being the human and ecological interests that are at the centre on the unsustainable trends and least well represented in the corridors of power.
(v) Sustainability assessment must respect the context. In all cases it must ensure integrated attention to the main overall requirements for reversing the trends and making progress towards sustainability. It must always aim to enhance the lasting integrity of social-ecological systems, deliver benefits to those lacking sufficiency and opportunity, multiply resource efficiency, protect the interests of future generations, favour precaution, build understanding and participative capacity. But it must in every application respect the particulars of the context, and specify the effective criteria for evaluations and decision making in light of the key issues and aspirations, capacities and concerns of the people and places involved.

(vi) Sustainability assessment must be, to the extent possible, open and broadly engaging. This is in part because sustainability assessment can be no mere technical exercise; it is always a matter of public choices among options and objectives for a desirable and lasting future. But openness and engagement are also needed because the challenges of building sustainability are beyond the capacities of governments and markets alone and consequently we must use all opportunities to foster the understanding and strengthen the participative capacities of citizens and civil society organizations.

There is more to it, of course. This introduction is only the beginning for a book that itself is also only a beginning. These are the early days of sustainability assessment and at some point, no doubt, we will look back in wonder about how primitive the thinking was back here at the outset. So be it. What matters now is that we accept the responsibility and opportunity to build, apply and improve an idea and instrument that is both remarkably positive and desperately needed.

References

Berra Y and Kaplan D (2002) *When You Come to a Fork in the Road, Take It!* (New York: Hyperion).

Butchart SHM, Walpole M, Collen B, Van Strien A, Scharlemann JPW, Almond REA, Baillie JEM, Bomhard B, Brown C, Bruno J, Carpenter KE, Carr GM, Chanson J, Chenery AM, Csirke J, Davidson NC, Dentener F, Foster M, Galli A, Galloway JN, Genovesi P, Gregory RD, Hockings M, Kapos V, Lamarque JF, Leverington F, Loh J, McGeoch MA, McRae L, Minasyan A, Morcillo MH, Oldfield TEE, Pauly D, Quader S, Revenga C, Sauer JR, Skolnik B, Spear D, Stanwell-Smith D, Stuart SN, Symes A, Tierney M, Tyrrell TD, Vié JC, Watson R (2010) 'Global biodiversity: indicators of recent declines', *Science* 328: 5982 (28 May), 1164–1168.

Diamond J (2005) *Collapse: How Societies Choose to Fail or Succeed* (New York: Viking).

FAO (UN Food and Agriculture Organization), Economic and Social Development Department (2008), *The State of Food Insecurity in the World, 2008: High Food Prices and Food Security – Threats and Opportunities* (Rome: FAO). [Online] Available at: <ftp://ftp.fao.org/docrep/fao/011/i0291e/i0291e00a.pdf> (accessed 10 June 2011).

Gibson RB (2006) *Sustainability-based Assessment Criteria and Associated Frameworks for Evaluations and Decisions: Theory, Practice and Implications for the Mackenzie Gas Project Review*, a report commissioned by the Joint Review Panel for the Mackenzie Gas Project. [Online] Available at: <http://www.acee-ceaa.gc.ca/default.asp?lang=En&n=155701CE-1> also <http://ssrn.com/abstract=1663015> (accessed 25 November 2011).

Gibson RB (2011) 'Application of a contribution to sustainability test by the Joint Review Panel for the Canadian Mackenzie Gas Project', *Impact Assessment and Project Appraisal* 29:3 (September), 231–244.

Gibson RB, Hassan S, Holtz S, Tansey J, Whitelaw G (2005) *Sustainability Assessment: Criteria and Processes* (London: Earthscan).

Glouberman S and Zimmerman B (2002) *Complicated and Complex Systems: What Would Successful Reform of Medicare Look Like? Discussion Paper #8, for the Commission on the Future of Health Care in Canada*. Available at: <http://www.change-ability.ca/publications/Entries/2002/7/1_Health_Care_Commission_-_Discussion_Paper_No.8.html> (accessed 10 June 2011).

Herrera AD, Scolnik HD, Chichilnisky G, Gallopin GC, Hardoy JE, Mosovich D, Oteiza E, de Romero Brest GL, Suarez CE, Talavera L (1976) *Catastrophe or New Society? A Latin American World Model* (Ottawa: IDRC).

IPCC (Intergovernmental Panel on Climate Change) (2007) *Climate Change 2007 Synthesis Report* (IPCC: Geneva). [Online] Available at: <http://www.ipcc.ch/publications_and_data/ar4/syr/en/contents.html> (accessed 14 June 2011).

IUCN (International Union for Conservation of Nature and Natural Resources) (1980) *World Conservation Strategy: Living Resource Conservation for Sustainable Development* (Geneva: IUCN-UNEP-WWF).

John G (2009) *Operation Imani: Ideas for Sustainable African Development*, documentary video (Ottawa: Canada Africa Community Health Alliance, 2010).

Malthus TR (1970 [1798]) *An Essay on the Principle of Population* (A Flew ed.) (Harmondsworth: Penguin).

Milanovic B (2005) *Worlds Apart: Measuring International and Global Inequality* (Princeton: Princeton University Press).

MEA (Millennium Ecosystem Assessment Board) (2005) *Current State and Trends Assessment*. [Online] Available at: <http://www.millenniumassessment.org/en/Condition.aspx> (accessed 10 June 2011).

Porritt J (2009) *Living Within Our Means*. Forum for the Future, 21 March 2009. [Online] Available at: <http://www.forumforthefuture.org/blog/living-within-our-means> (accessed 11 June 2011).

Ribaux S (2000) 'Warming trend: Quebec community groups are helping low-income households save energy and cut home heating bills', *Alternatives Journal* 26(2), 34.

UN (United Nations) (2010) *Millennium Development Goals Report 2010* (UN, New York). [Online] Available at: <http://www.un.org/millenniumgoals> (accessed 10 June 2011).

UNDP (United Nations Development Program) (2010) *Human Development Report 2010* (UNDP, Geneva). [Online] Available at: <http://hdr.undp.org/en/reports/global/hdr2010/> (accessed 10 June 2011).

Vaidyanathan G (2002) 'In Gandhi's footsteps: two unusual development organizations foster sustainable livelihoods in the villages of India', *Alternatives Journal* 28(2), 32–37.

WCED (World Commission on Environment and Development) GH Brundtland, chair (1987) *Our Common Future* (New York: Norton).

World Bank (2004) *Indicators*. [Online] Available at: <http://www.worldbank.org/data/wdi2004/> (accessed 10 June 2011).

WWF (2010) *Living Planet Report 2010* (WWF, Geneva). [Online] Available at: <http://wwf.panda.org/about_our_earth/all_publications/living_planet_report/> (accessed 10 June 2011).

Environmental assessment and sustainability timeline

1968 UNESCO conference on the rational use and conservation of the biosphere introduces ecologically sustainable development

1969 US *National Environmental Policy Act* introduces mandatory environmental assessment, with environment defined to include social, economic and cultural as well as biophysical aspects and their interrelations

1970s legislated as well as policy-based environmental assessment obligations are adopted in many industrialized countries; scholars and practitioners begin to develop approaches to linking social and biophysical assessment, considering cumulative effects, and facilitating more serious engagement of affected interests and citizen organizations

1972 UN Conference on Human Environment in Stockholm recognizes international environmental issues and presents the essential concept of combining environment and development

1972 Club of Rome publishes *Limits to Growth* (Meadows *et al.*, 1972)

1974 Bariloche Foundation team (Argentina) produces *Catastrophe or New Society?* on growth, equity and the environment (Herrera *et al.*, 1976)

1980s national environmental assessment regimes are established in poorer nations; scholars and practitioners develop the ecosystem approach to environmental assessment, expand attention to precaution in the context of uncertainty, and begin to explore iterative scenario-building and backcasting

1980 IUCN releases World Conservation Strategy, recognizing that conservation of nature depends on maintaining and enhancing the livelihoods of local people, explicitly promotes sustainable development (IUCN, 1980)

1984 the Consumer's Association of Penang, Malaysia, organizes the Third World Network to address international trade, development and environmental issues

1987 World Commission on Environment and Development releases *Our Common Future*, promoting 'sustainable development', which is quickly and widely adopted as an attractive alternative to both continuing ecological degradation and ultimate economic limits.

1987 World Health Organization introduces environmental health impact assessment integrating biophysical, social and psychological aspects

1988 Intergovernmental Panel on Climate Change is established to gather and assess biophysical and socio-economic research findings on climate change

1990s adoption of assessment requirements continues to expand globally but tensions between assessment requirements and commitments to minimally restricted economic growth increase in some jurisdictions; scholars and practitioners develop more advanced systems-based approaches to assessment, expand use of life-cycle analysis and give more attention to steps to integrate environmental assessment with land use, transportation and urban planning; requirements for strategic environmental assessment expand globally

1991 UN Economic Commission for Europe negotiates the *Espoo Convention* on environmental assessment of cases involving transboundary concerns

1992 World Business Council on Sustainable Development is founded

1992 UN Conference on Environment and Development in Rio de Janeiro produces Agenda 21, addressing a wide variety of matters including needs to eradicate poverty and cut consumption and production waste, national responsibilities to develop sustainable development strategies, adoption of the precautionary principle, promotion of local sustainability initiatives and engagement of a wide range of non-government participants

1998 UN Economic Commission for Europe adopts *Aarhus Convention* on Access to Information, Public Participation in Decision Making and Access to Justice in Environmental Matters

1999 Dow Jones introduces first global sustainability index for corporations listed on stock exchanges

2000s governments increasingly perceive limits to their own growth but face increasing pressures to attend to new problems and perceived entitlements, and to address evident vulnerabilities and deficiencies of financial systems and regulatory regimes; symbolic commitments to sustainability are gradually converted to more effective public expectations and mandatory requirements

2000 UN Millennium Summit sets commitments to Millennium Development Goals

2003 UN Economic Commission for Europe releases its *Protocol on Strategic Environmental Assessment*

2005 Millennium Ecosystem Assessment report reveals the human wellbeing implications of ecosystem services degradation (MEA, 2005)

2010 WWF's biennial *Living Planet Report* finds that human demands on nature's resources and services in 2007 had risen to 150 per cent of what the biosphere can sustain

2 The changing theory of impact assessment

Matthew Cashmore, Aalborg University
Lone Kørnøv, Aalborg University

Introduction

In the previous chapter, Robert Gibson provides a forceful analysis of the reasons why a diverse field of practice known as sustainability assessment has emerged. Scholars of sustainability assessment locate this concept within a broader heritage of policy instruments for improving decision making (Pope *et al.*, 2004). We label these as impact assessment (IA) instruments. In this chapter we aim to contextualise sustainability assessment into important social debates by examining how our theoretical understanding of IA has developed over time. The chapter concludes with reflections on the implications of these developments for contemporary scholarship on sustainability assessment, thereby linking historical endeavours with the content and innovations of this book.

The rapid economic and social development of the years following World War II, combined with social backlashes against concomitant negative effects, fostered rapid global institutionalisation of IA procedures from the 1970s onwards. IA was initially used to address the environmental effects of development projects, although in many places environment was interpreted broadly and included socio-economic dimensions of the human environment, albeit to varying degrees. IA has since morphed into multiple forms that address a broader range of issues (e.g. health impacts, regulatory burden, equity, gender equality, and, of course, sustainability assessment) and in a considerably extended range of decision contexts (e.g. nation state policy, international trade negotiations, and development co-operation). The proliferation of IA has also been perpetuated in recent years by the political emphasis on evidence-based decision making and the so-called knowledge society, leading to what has been aptly parodied as 'an analytical arms race' (Owens and Cowell, 2002: 50). Yet the popularity of IA, and its apparent pliability, provide substance for claims that it represents one of the most influential policy innovations of the twentieth century (Bartlett, 1988; Caldwell, 1998).

Whilst the practice of IA has burgeoned since the 1970s, theoretical issues have received comparatively modest attention. The need for greater consideration to be given to IA theory was repeatedly emphasised in the 1990s, as examinations of the breadth and diversity of theoretical deliberations occurring in other disciplines provided stark evidence of the introspective nature of IA scholarship (e.g. Lawrence, 1994; 1997). Such work also served to highlight a preponderance in IA scholarship with questions of how it should be practised – that is, its process and procedures – to the neglect of theoretical issues (Cashmore, 2004). Theory has certainly attracted increased attention in recent years, but nevertheless IA remains notably theoretically underdetermined.

In this chapter we seek to build an understanding of IA theory in order to contextualise discussions in subsequent chapters and to act, more generally, as a knowledge resource for scholarship on sustainability assessment. This is achieved by contrasting dominant beliefs in the 1970s and 1980s, which were typically expressed only implicitly as 'theories of practice', with the multivalency of contemporary theoretical influences and thoughts. The chapter commences with an introduction to the meaning of theory and an explanation of how scientific theory is interpreted herein. We then review early theoretical tendencies in IA scholarship to provide a critical introduction to its initial underpinnings. Following this, what we judge to be pivotal theoretical developments are examined; we focus on the three interlinked issues of the theoretical implications of changing normative beliefs, a changing theorisation of decision making, and developments in causal understanding. The issue of pluralism and what it means for theoretical scholarship is then considered, and the chapter concludes with reflections on what these developments might mean for contemporary scholarship on sustainability assessment.

Defining scientific theory

The term theory is widely used in both academic and everyday parlance, but it is a deceptively complex and much contested concept. It is partly for these reasons that the scientific literature contains a bewildering range of prefixes to qualify theories: for example, scientists talk about grand, meta- and middle-range theories; deductive, inductive and iterative theories; and critical and postmodern theories. There are also a number of closely related terms – notably conceptual framework and model (see Ostrom, 2007 for one interpretation of the distinction between these) – and oftentimes these terms are used loosely and interchangeably. In seeking to explain interpretations of scientific theory we start with the question of *why* theorise?

The policy arenas in which IA is employed, and of which it forms part, are invariably complex socio-cultural settings. They involve, amongst other things, multiple decisions and multiple actors with divergent interests, goals, values, and beliefs. In seeking to understand the operation of IA, and other social or natural phenomena, scientists simplify, using generalisations of varying degrees of specificity to reduce the 'manifold to simplicity' (James, 1959: 4). This practice reflects an important scientific principle: a belief that complex phenomena can be reduced to a smaller set of important relationships. Furthermore, the systematic description of theoretical premises is said to be a necessary precursor to validation, which is a central activity under most philosophies of science. Theory, within the IA research field and in modern science in general, becomes a tool of practice.

Turning to the issue of *what* is theory, Popper (2002: 37–38) defines it as the 'nets cast to catch what we call "the world": to rationalize, to explain and to master'. Many scientists use the term solely for presuppositions concerned with prediction and explanation – explanatory theory. Yet others use the term more broadly. For example, some identify descriptive theories (descriptions of the characteristics of a phenomenon, typically used when very little is known about the phenomenon under investigation) and relational theories (which focus on how dimensions of a phenomena are related). Our main focus in this chapter is on developments of, or pertinent to, explanatory theory.

Theoretical underpinnings in the early years

In order to understand recent developments in theoretical applications and concept development within the IA field, key dimensions of its early theoretical underpinnings are now considered. The professional practice of IA began in the US in 1970 as a consequence of the implementation of the National Environmental Policy Act (NEPA). In order to locate the theoretical underpinnings of IA we consider, firstly, the origins of IA as described in NEPA and subsequent related publications, and secondly, what was originally assumed about human decision making.

As previously mentioned, IA is rooted in the quest for a proactive approach to sustainable development in which environmental and other marginalised impacts could be assessed and incorporated into decision making. Through NEPA, IA institutionalised the thinking of trading off between environmental, social, and economic objectives (Wathern, 1988), with a fourfold purpose stated as:

> To declare a national policy which will encourage productive and enjoyable harmony between man and his environment; to promote efforts which will prevent or eliminate damage to the environment and biosphere and stimulate the health and welfare of man; to enrich the understanding of the ecological systems and natural resources important to the Nation; and to establish a Council on Environmental Quality.
>
> (Public Law 91-190, 42 U.S.C. 4321)

Lynton Caldwell, a principal architect of NEPA, describes IA as a 'mandatory, continuing, systematic, integrated, science-based policy analysis' (1982: 51) and emphasises its value in helping to 'raise the national consciousness concerning the environment' and 'elevate attitudes and action' (1998: 6). That NEPA is grounded in both substantive (subject to values and goals) and procedural (formal) rationality is argued by several authors (Bartlett, 1986; Caldwell, 1982; Boggs, 1995), and according to these arguments the focus in IA is on both *how* IA should do things and *where* to go based upon the assessments.

In parallel, other interpretations of the philosophy and principles of NEPA exist, in which IA is traced back to a procedural rationalistic approach to decision making (Jay *et al.*, 2007; Owens *et al.*, 2004; Kørnøv and Thissen, 2000). The arguments here are based upon an understanding of IA as an expert system – an example of a procedurally rational decision-making approach consisting of a number of steps and rules that specify behaviour. IA is thereby based upon a simple idea about the relationship between information and decision making: by providing more and 'better' (i.e. scientific, reliable, quantified, etc.) environmental knowledge, a 'better' or more rational decision, defined as 'one that pursues a logic of consequences' (March, 1994: 2), will be reached. It can further be argued that IA procedures in general do not force decision makers to select more environmentally friendly alternatives; invariably, the emphasis is on procedural rationality. This is confirmed by legal decisions in the US, and training manuals and agency guidelines that 'focus on procedural requirements, rather than offering any specific guidance for decision making' (Stern and Mortimer, 2009: 14).

Despite being envisaged as having a broad-ranging purpose, the theoretical underpinning of IA is much influenced by procedural rational decision making as a normative ideal. This is connected to theoretical developments which took place towards the end of the Second World War, during which time theories on rational decision making (such as modern utility theory) appeared and became established. IA scholars adopted these economic and mathematical theories of decision making and, based on them, formed principles, techniques, and approaches for IA. Expected utility theory, by Von Neumann and Morgenstern (1944), assumes that decision makers tend to choose among alternatives with probabilistic consequences to maximise utility, or in the case of IA a maximisation of environmental management ideals. The maximisation is based upon choices consistent with an individual's preferences. This utilitarian viewpoint has had a significant impact on IA, but has also been challenged, not least due to a fundamental theoretical limit: future societies' preferences cannot be known.

Since its early theoretical development, scholarship on IA has become increasingly diverse in terms of such factors as disciplinary inputs, understandings of decision making, and philosophical positioning. It is suggested that four main approaches to, or schools of thought on, IA can be identified since its inception in the 1970s:

1 *Rational IA decision-making approaches.* IA is interpreted as a systematic procedure with steps that actors can or should take in order to reach a

decision that balances environmental and economic goals. This approach implies a conscious social actor engaging deliberately in (calculative) assessment strategies. This approach prevailed during the period between 1970 and 1985.

2. *Descriptive models of IA and decision making.* Through implementation and empirical testing it became apparent that IA actors do not always undertake and use assessments in rational ways. The descriptive approach was compelling from 1985 to 1995, as is evident in the numerous empirical case studies of IA in practice. This approach provides the basis for what can be termed a 'theory of practice' prescribing what needs to be done and sometimes why.
3. *Institutional approaches to IA.* The focus moves under this approach from 'what ought to be' to 'real' IA processes. Starting around 1995, this approach emphasises institutional setting and contextual complexity. Scholarship draws extensively upon theories from other disciplines, such as organisational and decision-making theory, learning theory and policy science, to shed light on context dependency in IA.
4. *Constructivist approaches to theorising IA.* Constructivism is a position within philosophies of science in which knowledge of the world is viewed as socially constructed and hence indeterminate. It has become increasingly influential within the social sciences since the so-called science wars of the 1990s, and is strongly reflected in contemporary scholarship on IA. The directions in which constructivist research is moving IA scholarship are many and diverse, with theoretical implications which are very much unresolved at present.

The theoretical underpinning in the first period has been described in this section, and the second (descriptive models of IA) are covered extensively elsewhere. Our discussion of the changing nature of IA theory and pluralism in the following sections draws primarily on the third and fourth categories of IA described above.

The changing nature of IA theory

Given that IA has been the subject of considerable scholarly interest for some four decades, it is unremarkable that our understanding of it has changed over this time. The most marked developments in our understanding, however, are relatively recent phenomena, confined mainly to the last 15 years. These 'recent' innovations reflect, in part, a new found willingness to engage with scholarly discourses outside the immediate purview of IA practices. Research on, *inter alia*, the use (and non-use) of evidence in policy making, evaluation, and organisational learning have provided fertile grounds for stimulating theoretical developments. The critical turn in the social sciences has also encouraged the problematisation of received truths and norms, leading to the introduction of rich new perspectives that diverge sharply from conventional thinking.

It is not possible in this short chapter to cover the entire spectrum of theoretical developments in IA scholarship, nor is it feasible to enter into a detailed examination of the many nuances of particular theorisations. Instead, we have selected three interrelated issues to illustrate what are, in our opinion, important focal points of theoretical work and hence central to developments concerning sustainability assessment. We commence by considering significant normative developments that have taken place since the 1970s and their effects upon IA scholarship. Developments in decision-making theory and their implications for the theorisation of IA are then examined. Finally, we draw together these two focal points of scholarship in considering causal theory.

The changing normative basis for IA theory

Normative beliefs prevalent in a particular culture provide a framework defining the legitimate and purposeful use of IA tools and, as such, have important implications for theory. It is axiomatic that normative beliefs have altered since IA was first institutionalised. The single most important normative shift concerns the adoption of sustainable development internationally as a key goal of human development. This has led to a repositioning of IA, with sustainable development now invariably seen as its primary purpose. This is a sinuous development, for positive interpretations of the concept of sustainable development vary considerably in terms of their normative prescriptions (see Chapter 3). But it has served to foreground in IA scholarship issues of equity, differentiated needs and responsibilities, limits to growth, and precaution.

The notion of sustainable development has also been a prominent factor driving changes in governance principles, which has occurred concurrently with fairly radical changes to the power and authority of sovereign governments. Like sustainable development, governance is a contested concept, the meaning of which is continually reconstructed. It is generally accepted, however, that governance for sustainable development emphasises participation and dialogue (as distinguished from consultation), transparency, accountability, and justice. Expectations concerning the appropriate role and use of experts in governance have also been recast as public trust in science has been shaken, plus there is greater recognition of the value of incorporating alternative forms of knowing (variously described as local, Indigenous, and tacit knowledge) into decision making (Bond *et al.*, 2010). Thus, Owens *et al.* (2004: 1948) suggest that, '[f]or a growing number of commentators, the human capacity for judgement, and especially for reaching intersubjective judgements through a process of argument and debate – is an asset to be nurtured'.

These two important normative changes self-evidently have significant implication for our understanding of IA. Repositioning of its purpose from the narrow expectation of informing decisions to contributing to sustainable development has transformed theory about how IA can be purposefully applied. One manifestation of this is scholarship on using IA to engender learning (e.g. Jha-Thakur *et al.*, 2009; Hertin *et al.*, 2007). It has also opened up for theoretical investigation issues concerning the analytics of change in complex, interlinked ecological and social systems, at a number of scales (Bond and Morrison-Saunders, 2011). Similarly, changing governance expectations have fostered theoretical work on how IA can generate civil legitimacy under the revised 'rules of the game' and in ways that are cognisant of culture. Work has focused particularly on building civil legitimacy through interactive and reflexive dialogical processes, and on understanding the circumstances in which particular knowledges – scientific or otherwise – are viewed as pertinent and useful.

Developments in theories of decision making

Decision theory concerns how individuals and groups make decisions, and is intended both to explain and to predict human behaviour. The different theories can be categorised into:

- normative theories – explaining how decisions ought to be taken.
- descriptive theories – explaining how decisions are actually taken.
- prescriptive theories – seeking to improve decision making by addressing the limitations identified in descriptive theories.

In the development of decision-making theories there is a general trend from normative to descriptive and prescriptive theories, and the portrayal of decision making has developed from unbounded to bounded rationality, from de-contextualised to context dependent, and from rational reasoning to a critical perspective on power and ambiguity. These developments are now exemplified.

Stemming largely from economics and statistics, the classic rational decision-making theory assumes an economic decision maker who is fully informed regarding possible options and outcomes of options, infinitely sensitive to the distinctions among options, and fully rational regarding their choice of options (Sternberg, 2009). By the mid 1950s the theory of bounded rationality was proposed by Simon, who advocated an alternative model – the administrative model – of decision making. The administrative model differed from the classic model in a number of key respects: it replaced 'the goal of maximizing with the goal of satisficing' (Simon, 1957: 204); it divided decision-making responsibilities between specialists; and it incorporated sub goals in addition to global goals (Simon, 1979). Simon's introduction of a descriptive theory of bounded rationality has been valuable to understanding real-world decision-making processes, and in addition to the cognitive limitation, Simon also pointed to the need for a greater understanding of the decision context while 'individual choice takes place in an environment of "givens"' (Simon, 1957:

201). Later, by reformulating it as a 'bounding of rationality', Forester opens up this discussion and raises questions like 'what bounds rational actions?, how does such bounding take place?' (Forester, 1993: 74) and thereby focuses on the social structures and specific decision contexts a decision maker is confronted with. In line with Forester, March focuses on context dependency and develops a descriptive theory of rule following, in which the logic of consequences is replaced by a logic of appropriateness, with socially constructed rules playing a decisive role (March, 1994). Alongside theoretical developments with a critique of the rational assumption, the political model of decision making was developed. This model emphasises that decisions follow preferences and that choices reflect the preferences of the most powerful (March, 1962; Salancik and Pfeffer, 1974). As such, the political model emphasises such characteristics of decision making as control of agenda and access, the formation of coalitions, and lobbying activities (Pettigrew, 1973; Pfeffer, 1992).

The rational, administrative, and political models of decision making are criticised by Cohen and colleagues for their lack of sensitivity to the complex, unstable, and ambiguous context in which decision making takes place (Cohen *et al.*, 1972). Their view of decision making as an unstructured process is taken to the extreme in the garbage can model, in which decisions are the result of random interactions of problems, solutions, participants, and choice opportunities (Cohen *et al.*, 1972). The role of ambiguity is also underlined in the theoretical discussion on 'taste ambiguity' by March describing how preferences are managed, constructed, avoided, expected to change, treated strategically, suppressed, and confounded in the decision-making process (March, 1978).

These examples of shifts within decision theory point to complexity as an intrinsic property of decision making. Decision making is influenced by cognitive limits, the social context with goal disagreement and non-consistent and changing preferences.

The de-contextualised ideal embedded in the rational choice perspective has also influenced IA, which from the beginning was treated as a broadly rational activity. Even though IA legislation and guidelines to a large extent still use the language of, and present assessment procedures in line with, the rational paradigm of decision making, the general understanding is that this can scarcely claim to describe the realities around IA. The use of descriptive decision-oriented theory has increased, especially over the last decade (e.g. Kørnøv and Thissen, 2000; Nilsson and Dalkmann, 2001; Pischke and Cashmore, 2006; Bond, 2003), based in part on an argument that an increase in the realism of both the political and cognitive underpinnings of IA will improve the field and practice by generating theoretical insight. This has resulted in prescriptive models like ANSEA (Analytical Strategic Environmental Assessment: see Caratti *et al.*, 2004). The political model of decision making is also being brought into IA research. For example, Richardson (2005) underlines the need for IA to engage with competing values and suggests that IA itself is an instrument, or camouflage, of power.

The complexity in, and of, contexts in which IA is being developed, implemented, and evaluated has recently received more attention. Systemic complexity can be analysed with first- and second-order theoretical perspectives: first order with a focus on the system itself and second order considering who is describing the system and how (Tsoukas and Hatch, 2001). Examples of first-order perspectives include work on the organisational and institutional context of IA and decision making (Bina, 2007; Bina, 2008; Hilding-Rydevik and Bjarnadóttir, 2007) and on understanding policy processes (Kørnøv and Thissen, 2000; Nitz and Brown, 2001). That understanding of complexity is observer-dependent (Casti, 1986) in a second-order theory is beginning to be used in IA literature. For example, Weick's sense-making theory (1995) is used to illustrate how IA, and our understanding of it, is constructed through interactive sense-making processes between actors in the system and the describer of the system (Lyhne, 2011).

Causal theory

We now turn to consider how aspects of the developments discussed in the previous two sub-sections – on embracing complexity, cultural sensitivity, and the need for civil legitimacy, the political model of decision making, etc. – have cumulatively affected scholarship on a central component of explanatory theory: causal thinking. A seminal contribution to the redirection of causal theory in IA was made by Bartlett and Kurian (1999), who delineated six models of the causal operation of IA from the research literature (see Table 2.1). Their work – whilst presenting a synopsis of how IA *is* understood, rather than how it *could be* understood – helped greatly to structure subsequent theoretical and empirical research by highlighting the key assumptions embedded in IA scholarship. What is intriguing about the models is that they each have explanatory potential. Even the information processing model (akin to what we have labelled as rational IA approaches), which has been subjected to trenchant critique over several decades, may have explanatory value in limited circumstances (Owens *et al.*, 2004). In consequence, causation is generally seen to involve multiple, possibly complimentary pathways to effecting sustainable development.

Nevertheless, of the six models Bartlett and Kurian delineated, much recent theoretical work has focused on the institutional one, for this has been viewed as having arguably the greatest potential for instigating the types of long lasting and far reaching societal changes necessary for sustainable development. A key notion in institutional theory is how resources (e.g. staff capacity) and learning can influence formal and informal rules in the long term by altering the values, norms, and maybe even world-views, which underpin them. A number of important research themes have emerged in relation to this. These include scholarship on:

Table 2.1 Models of the causal operation of IA (based on Bartlett and Kurian, 1999)

Model	Causal assumptions
Information processing	IA produces information for use in apolitical decision making.
Political economy	IA forces markets to internalise particular (e.g. environmental, health, social, etc.) policy externalities.
Symbolic politics	IA operates by evoking emotional responses and the reaffirmation of moral commitments to particular policy issues based on rationalities that could be variously interpreted either as just or unjust.
Pluralist politics	Impact occurs by opening up decisions to the polity.
Organisational politics	Change is a product of reform of the ways in which decisions are made internally within organisations.
Institutional	Formal and informal rules operating in the political administration are recast as a result of altered values, beliefs and world-views.

- The relationship between institutional influence and the level of integration of IA with decision making (e.g. Dalkmann *et al.*, 2004; Stoeglehner *et al.*, 2009).
- Barriers to achieving institutional influence, including capacity requirements for IA (e.g. Turnpenny *et al.*, 2008; Kolhoff *et al.*, 2009).
- The effect of culture – political, organisational, administrative, and legal – on institutional influence (e.g. Zhu and Ru, 2008; Jha-Thakur *et al.*, 2009).

The political naivety of much writing on IA has been remarked upon by several commentators and made explicit, at least to a certain extent, in political models of decision making. Recent work has consequently placed greater emphasis on the theoretical implications for causal understanding of both the political nature of the arenas in which IA is employed and the political character of IA instruments themselves. A fascinating example of this 'political turn' is the compelling work of Bent Flyvbjerg (1998) on planning, rationality, and power. For Flyvbjerg, IA is an example of policy instruments that are 'deeply embedded in the hidden exercise of power and the protection of special interests' and implicated in 'defining reality' (Flyvbjerg, 1998: 225 and 227, respectively). A rather more mundane, but nevertheless theoretically enlightening, picture of the politics of IA is presented in Cashmore *et al.* (2008) in their reflections on the political mediation of causality in IA practices. They meticulously chart the use of IA in essentially an advocacy manner, through the employment of various conciliatory mechanisms, to achieve development proponents' primary goal of achieving development consent.

Flyvbjerg has done much to raise awareness about power considerations in IA, but his headline conclusion that 'planners and promoters deliberately misrepresent costs, benefits, and risks' (Flyvbjerg, 2007: 578) may be overly simplistic. Whilst advocacy and strategic representation are undoubtedly prolific, other work on power paints a considerably more complex picture of the ways in which it permeates IA theory and practice (e.g. Cashmore *et al.*, 2010; Richardson and Cashmore, 2011). It may involve, for example, effects arising as a consequence of the subconscious reproduction of essentially unquestioned social norms. In particular, such work emphasises the subjectivity of IA: for example, why is 'good governance' deemed good; how is legitimacy for new forms of, or approaches to, IA constructed; and when is the exercise of power legitimate? The implications of such work for causal theory are unclear, but it evidently problematises scholarly understanding of how IA can, and is, used. We now consider other dimensions of the implications of the politicisation of causal theory as part of our examination of theory and plurality.

Theory and pluralism

The issue of pluralism is a central concern of this book. Within the context of IA theory, pluralism has several important connotations. We address two particularly salient dimensions of plurality for IA scholarship: firstly, the plurality of theoretical perspectives and secondly, the implications of constructivism, and its emphasis on indeterminacy and hence plurality of stakeholders in any IA setting with multiple perspectives of what is good, appropriate, or sustainable in decision making.

It is evident from the previous sections that a range of theoretical perspectives exists concerning the operation of IA. It was noted previously that for most philosophies of science the testing of theories is fundamental to developing credible knowledge. The existence of multiple theories could then be interpreted as evidence of the need for greater emphasis to be placed on empirical research. There are certainly strong elements of truth in this, for too much debate on IA has been poorly grounded in empirical data, although conversely, too much research has constituted naive empiricism lacking any obvious connection to theory. Yet there are a number of methodological challenges (notably attribution and timescale – see Chapter 4) associated with testing theories that apply to dynamic social settings. The implication is that theoretical indeterminacy is likely to remain a key feature of IA for the foreseeable future. Although those searching for policy prescriptions might view indeterminacy as frustrating, it can also be interpreted as a sign of an actively developing, vibrant and critical field of enquiry.

The second issue concerns the ongoing transition in the social sciences from realism to constructivism (Jasanoff, 2004). This is significant for theoreticians because constructivists' view of knowledge as indeterminate raises a number of intriguing questions, particularly when coupled with the normative belief in pluralist politics prevalent in Western democracies. There are theoretical issues about how the operation of IA is understood: if science is no more or less fallible than any other form of knowing, then what does constructivism mean for our understanding of the recreation of power in and through IA systems? We concur with Owen and Cowell's (2002: 53) conclusion that one of the significant strengths of IA is that it 'can mobilise different conceptions of sustainability, depending upon the context within which it is applied and what precisely is asked of the technique'. This applies equally whether sustainability assessment or one of the many other types of IA is being considered. But if multiple expectations concerning IA are simply alternative social constructs (and hence equally legitimate) and occur concurrently in a given society, what does this mean for how IA is conceptualised and its societal role interpreted?

In an applied discipline like IA the implications of constructivist beliefs for the theory–practice nexus should unashamedly be given high priority. Constructivism problematises the hitherto dominant perception that practical recommendations can be derived straightforwardly from theory; or to put it another way, that a universal procedural heuristic is grounded in theory. As mentioned in the previous section, an alternative conception is gaining ground, wherein theory is viewed more as a rich body of knowledge which is used reflexively to develop practices that build civil legitimacy over the purposes of IA, the means by which these are achieved, and the issues that are addressed (Cashmore *et al.*, 2008; Bond *et al.*, 2011).

Conclusions

Our aim in this chapter was not to delineate a particular field of theoretical activity, which can be labelled as pertaining to, or even constituting, sustainability assessment. To do so, would place artificial boundaries around one aspect of a broad and diverse field of endeavour, and thereby run the risk of encouraging introspection and parochialism where a broad awareness of contemporary scholarship, plus open-mindedness are required. Our more modest goal has been to contextualise, from a theoretical perspective, key discussions which take place in the following chapters.

Sustainability assessment is very much an endeavour rooted in the historical development of IA and there is much that can be learnt from this, and other, applied fields of practice. After all, Bond and Morrison-Saunders' (2011) interpretation of sustainability assessment as a dialogical vehicle for defining sustainability in its context might not be too different from how environmental impact assessment would be conceptualised if it were being implemented today (see, for example, Elling, 2008). Furthermore, the degree of theoretical differentiation between IA types is overstated, purposefully or otherwise, in much of the literature; there is considerably more common theoretical ground than research and practice would often seem to imply.

Building on our examination of the changing nature of IA theory, we conclude by identifying three critical avenues for scholarly work on sustainability assessment:

- *Reason, or the philosophical positioning of sustainability assessment.* The issue of reason in IA (or what is often described as its rationality), and of governance and society more generally, remain at the forefront of theoretical work, inseparable from many, if not most, theoretical concerns. This is taken up in the following chapters in debate on, amongst other things, the meanings of sustainable development and effectiveness.
- *Sustainability transitions, as an aspect of causal theory.* Our understanding of the contexts under which various forms of change take place, and their potential consequences, remains poorly developed. Within the context of sustainability assessment, there is the additional complicating factor that the changes involved in a transition to sustainability are generally conceived as being more profound in nature than has tended to be the case for other forms of IA. This is addressed in the following chapters through a focus upon issues of conceptual and analytical boundaries (for example, in terms of time and space), and learning.
- *Ethical dimensions of sustainability assessment.* The ethical dimensions of IA instruments intended for use in highly politicised arenas, for specific governance objectives (e.g. intergenerational equity, a focus on the needs of the poor, better quality of life, etc.), and where there is explicit recognition of the legitimacy of plurality are critically important, but poorly theorised to date. This is taken up in the following chapters in debates pertaining to the treatment of norms and values in the evaluation of effectiveness.

References

Bartlett RV (1986) 'Rationality and the logic of the National Environmental Policy Act', *The Environmental Professional*, 8, 105–111.

Bartlett RV (1988) 'Policy and impact assessment: an introduction', *Impact Assessment Bulletin*, 6, 73–74.

Bartlett RV and Kurian PA (1999) 'The theory of environmental impact assessment: implicit models of policy making', *Policy and Politics*, 27, 415–434.

Bina O (2007) 'A critical review of the dominant lines of argumentation on the need for strategic environmental assessment', *Environmental Impact Assessment Review*, 27, 585–606.

Bina O (2008) 'Context and systems: thinking more broadly about effectiveness in strategic environmental assessment in China', *Environmental Management*, 42, 717–733.

Boggs JP (1995) 'Procedural vs. substantive in NEPA law: cutting the Gordian Knot'. In: Lemons J (ed.) *Readings from The Environmental Professional* (Oxford: Blackwell Science).

Bond A, Dockerty T, Lovett A, Riche AB, Haughton AJ, Bohan DA, Sage RB, Shield IF, Finch JW, Turner MM, Karp A (2011) 'Learning how to deal with values, frames and governance in sustainability appraisal', *Regional Studies*, 45(8), 1157–1170.

Bond A and Morrison-Saunders A (2011) 'Re-evaluating sustainability appraisal: aligning the vision and the practice', *Environmental Impact Assessment Review*, 31, 1–7.

Bond AJ (2003) 'Let's not be rational about this', *Impact Assessment and Project Appraisal*, 21, 266–269.

Bond AJ, Viegas CV, Coelho CCSR, Selig PM (2010) 'Informal knowledge processes: the underpinning for sustainability outcomes in EIA?', *Journal of Cleaner Production*, 18(1), 6–13.

Caldwell LK (1982) *Science and the National Environmental Policy Act* (Tuscaloosa: University of Alabama Press).

Caldwell LK (1998) 'Implementing policy through procedure: impact assessment and the National Environmental Policy Act (NEPA)'. In: Porter, AL and Fittipaldi, JJ (eds) *Environmental Methods Review: Retooling Impact Assessment for the New Century* (Fargo, ND: The Press Club).

Caratti P, Dalkmann H, Jiliberto R (eds) (2004) *Analysing Strategic Environmental Assessment: Towards Better Decision Making* (Cheltenham: Edward Elgar).

Cashmore M (2004) 'The role of science in environmental impact assessment: process and procedures versus purpose in the development of theory', *Environmental Impact Assessment Review*, 24, 403–426.

Cashmore M, Bond A, Cobb D (2008) 'The role and functioning of environmental assessment: theoretical reflections upon an empirical investigation of causation', *Journal of Environmental Management*, 88, 1233–1248.

Cashmore M, Richardson T, Hilding-Ryedvik T, Emmelin L (2010) 'Evaluating the effectiveness of impact assessment instruments: theorising the nature and implications of their political constitution', *Environmental Impact Assessment Review*, 30, 371–379.

Casti J (1986) 'On system complexity: identification, measurement, and management'. In: Casti J and Karlqvist A (eds) *Complexity, Language, and Life: Mathematical Approaches* (Berlin: Springer-Verlag).

Cohen MD, March JG, Olsen JP (1972) 'A garbage can model of organizational choice', *Administrative Science Quarterly*, 17, 1–25.

Dalkmann H, Jiliberto R, Bongardt D (2004) 'Analytical strategic environmental assessment (ANSEA) developing a new approach to SEA', *Environmental Impact Assessment Review*, 24, 385–402.

Elling B (2008) *Rationality and the Environment: Decision Making in Environmental Politics and Assessment* (London: Earthscan).

Flyvbjerg B (1998) *Rationality and Power: Democracy in Practice* (London: University of Chicago Press).

Flyvbjerg B (2007) 'Policy and planning for large-infrastructure projects: problems, causes, cures', *Environmental and Planning B: Planning and Design*, 34, 578–597.

Forester J (1993) *Critical Theory, Public Policy, and Planning Practice: Towards a Critical Pragmatism* (New York: State University of New York Press).

Hertin J, Turnpenny J, Jordan A, Nilsson M, Russel D, Nykvist B (2007) 'Rationalising the policy mess? Ex ante policy assessment and the utilisation of knowledge in the policy process', *Environment and Planning A*, 41, 1185–120.

Hilding-Rydevik T and Bjarnadóttir H (2007) 'Context awareness and sensitivity in SEA implementation', *Environmental Impact Assessment Review*, 27(7), 666–684.

James W (1959) *Essays in Pragmatism* (New York: Hafner Publishing).

Jasanoff S (2004) *Design on Nature: Science and Democracy in Europe and the United States* (Princeton: Princeton University Press).

Jay S, Jones CE, Slinn P, Wood C (2007) 'Environmental assessment: retrospect and prospect', *Environmental Impact Assessment Review*, 27, 287–300.

Jha-Thakur U, Gazzola P, Peel D, Fischer TB, Kidd S (2009) 'Effectiveness of strategic environmental assessment: the significance of learning', *Impact Assessment and Project Appraisal*, 27, 133–144.

Kolhoff AJ, Runhaar HAC, Driessen PPJ (2009) 'The contribution of capacities and context to EIA system performance and effectiveness in developing countries: towards a better understanding', *Impact Assessment and Project Appraisal*, 27, 271–282.

Kørnøv L and Thissen WAH (2000) 'Rationality in decision- and policy-making: implications for strategic environmental assessment', *Impact Assessment and Project Appraisal*, 18, 191–200.

Lawrence DP (1994) 'Designing and adapting the EIA planning process', *The Environmental Professional*, 16, 2–21.

Lawrence DP (1997) 'The need for EIA theory-building', *Environmental Impact Assessment Review*, 17, 79–107.

Lyhne, I (2010) *Strategic Environmental Assessment & The Danish Energy Sector. Exploring Non-Programmed Strategic Decisions*. Doctoral Thesis. (Aalborg University: The Danish Centre for Environmental Assessment, Department of Development and Planning).

March JG (1962) 'The business firm as a political coalition', *Journal of Politics*, 24, 662–678.

March JG (1978) 'Bounded rationality, ambiguity, and the engineering of choice', *Bell Journal of Economics*, 9, 587–608.

March JG (1994) *A Primer on Decision Making: How Decisions Happen* (New York: The Free Press).

Nilsson M and Dalkmann H (2001) 'Decision making and strategic environmental assessment', *Journal of Environmental Assessment Policy and Management*, 3, 305–327.

Nitz T and Brown AL (2001) 'SEA must learn how policy making works', *Journal of Environmental Assessment Policy and Management*, 3, 329–342.

Ostrom E (2007) 'Institutional rational choice: an assessment of the institutional analysis and development framework'. In: Sabatier PA (ed.) *Theories of the Policy Process*. Second edn. (Boulder: Westview Press).

Owens S and Cowell R (2002) *Land and Limits: Interpreting Sustainability in the Planning Process* (London: Routledge).

Owens S, Rayner T, Bina O (2004) 'New agendas for appraisal: reflections on theory, practice, and research', *Environment and Planning A*, 36, 1943–1959.

Pettigrew AM (1973) *The Politics of Organizational Decision Making* (London: Tavistock).

Pfeffer J (1992) *Managing With Power: Politics and Influence in Organizations* (Boston: Harvard Business School Press).

Pischke F and Cashmore M (2006) 'Decision-oriented environmental assessment: an empirical study of theory and methods', *Environmental Impact Assessment Review*, 26, 643–662.
Pope J, Annandale D, Morrison-Saunders A (2004) 'Conceptualising sustainability assessment', *Environmental Impact Assessment Review*, 24, 595–615.
Popper K (2002) *The Logic of Scientific Discovery* (London: Routledge Classics).
Richardson T (2005) 'Environmental assessment and planning theory: four short stories about power, multiple rationality and ethics', *Environmental Impact Assessment Review*, 25, 341–365.
Richardson T and Cashmore M (2011) 'Power, knowledge and environmental assessment: the World Bank's pursuit of good governance', *Journal of Political Power*, 4 (1), 105–125.
Salancik GR and Pfeffer J (1974) 'The bases and use of power in organizational decision making: the case of a university', *Administrative Science Quarterly*, 19(4), 453–473.
Simon H (1957) *Administrative Behaviour* (London, MacMillan).
Simon H (1979) 'Rational decision making in business organisations', *American Economic Review*, 69(4), 493–513.
Stern MJ and Mortimer MJ (2009) *Exploring National Environmental Policy Act Processes Across Federal Land Management Agencies* (Washington DC: United States Department of Agriculture, Forest Service).
Sternberg RJ (2009) *Cognitive Psychology* (Belmont: Wadsworth).
Stoeglehner G, Brown AL, Kørnøv LB (2009) 'SEA and planning: "ownership" of SEA by the planners is the key to its effectiveness', *Impact Assessment and Project Appraisal*, 27, 111–120.
Tsoukas H and Hatch MJ (2001) 'Complex thinking, complex practice: the case for a narrative approach to organizational complexity', *Human Relations*, 54, 979–1013.
Turnpenny J, Nilsson M, Russel D, Jordan A, Hertin J, Nykvist B (2008) 'Why is integrating policy assessments so hard? A comparative analysis of the institutional capacities and constraints', *Journal of Environmental Planning and Management*, 51, 759–775.
Von Neumann J and Morgenstern O (1944) *Theory of Games and Economic Behavior* (Princeton: Princeton University Press).
Wathern P (ed.) (1988) *Environmental Impact Assessment: Theory and Practice* (New York: Routledge).
Weick, KE (1995) *Sense-making in Organizations* (US: SAGE Publications).
Zhu D and Ru J (2008) 'Strategic environmental assessment in China: motivations, politics, and effectiveness', *Journal of Environmental Management*, 88, 615–626.

Part 2

Sustainability assessment: pluralism

3 Challenges in determining the effectiveness of sustainability assessment

Alan Bond, University of East Anglia
Angus Morrison-Saunders, Murdoch University and North West University

Introduction

Considering the environment (including the place of human beings in it), and how best it might be managed, is really thinking about environmental governance. As an approach to environmental governance, sustainable development is just one of many 'discourses'[1] which exist and we have previously made the point that it reflects a view that socio-economic development and environmental conservation are, to an extent, compatible goals and that socio-economic development is necessary (Bond and Morrison-Saunders, 2009). Not everyone shares this view and it is important to bear in mind that whilst this book assumes sustainable development is a good thing, proponents of 'deep ecology' (as just one example) would argue that sustainable development inappropriately favours an anthropocentric view of the world in which humans have a right to dominate nature (Grey, 1993; Jacob, 1994; Williams and Millington, 2004).

Sustainability assessment is based on an implicit premise that sustainable development is the appropriate discourse on environmental governance. The fact that there are other discourses on environmental governance has implications for whether sustainability assessment can ever be considered 'effective' as it is promoting a governance view which is not universally held. That said, sustainable development 'has become the dominant rhetorical device of environmental governance' (Adger *et al.*, 2003: 1095) and, therefore, could be argued to be the dominant discourse in environmental decision making in most jurisdictions at present. In this context, sustainability assessment needs to be contributing to the achievement of sustainable development, and it is based on this assumption that this chapter and the overall book is written.

If sustainability assessment is to be of lasting value in shifting institutional decision making towards less environmental and social damage and more just and lasting relations between natural and human systems, its effectiveness needs to be considered in terms of practice. This chapter identifies key issues that need to be considered when attempting to establish just how 'effectiveness' might be judged, and to set the scene for chapters that follow. In this book, we do not prescribe a singular and authoritative definition of sustainability assessment and what it should and should not involve, but some statement is warranted. Sustainability assessment is, to put it simply, a process that directs decision making towards sustainability; it can be applied to any field of endeavour and any type of decision (derived from Hacking and Guthrie, 2008).

Whether a process like sustainability assessment delivers on its promise of increased sustainability depends a lot on the values and aspirations that frame the questioning. In any particular country, there will be legislative and procedural constraints that prescribe how sustainability assessment is undertaken and what happens to the information and advice it produces, but there will also be diverse values and opinions that provide foundations for political judgement about its effectiveness and what it delivers. This mirrors the case for Environmental Impact Assessment (EIA) where Fuller (1999) set out differing key stakeholder expectations as, amongst others, for proponents, certainty of outcome; for the public, the right to be heard; and for decision makers, a minimisation of delays.

This chapter sets out three key challenges that need to be considered when attempting to establish just how 'effectiveness' might be judged. In doing so, it sets the baseline for Chapter 8 to develop a framework for evaluating effectiveness that is put to use in Chapters 9 through to 12 which describe and evaluate sustainability assessment as it is practiced in four very different jurisdictions.

Initially, our first challenge is to consider what might be contested, that is, the issues which are critical to sustainability assessment, but which have no single, accepted meaning or definition. We refer to these as critical debates in that they are central to the interpretation of sustainability assessment, and also there tends to be a spectrum of views between (often) polar opposites within each particular debate. Our discussion indicates just why sustainability assessment can deliver outcomes which are variously interpreted as being good or bad (effective or ineffective) by different stakeholders – even before we consider the questions of temporal and spatial scales in Chapters 4 and 5 and legal and social pluralism in Chapters 6 and 7. Our discussion also serves to provide some guidance to practitioners for how to design a sustainability assessment process.

Our second challenge is to consider how 'effectiveness' might be measured. Some considerable progress has been made by researchers in terms of categorising effectiveness and developing criteria and this work will be summarised. The third challenge relates to the issue of pluralism and the implications it has for the interpretation of effectiveness; a particular challenge being that effectiveness will be measured based on a framing that favours some views over others.

One of the key lessons from the chapter is that the criteria that have been developed to date do not fully address the critical debates and fail to accommodate pluralism, and this point is considered in the conclusions.

Critical debates

Below, we examine five critical debates that determine the trajectory of sustainability assessment in practice. While we are sure there are more debates that we are not yet aware of, experience to date has exposed these five as especially deserving of attention by practitioners when designing and undertaking a sustainability assessment.

Contested meanings of sustainable development

Definitions of sustainable development abound and, at least on the surface, it is easy to blithely define the concept. Drawing on the terms developed in the original Brundtland Report (World Commission on Environment and Development, 1987), most definitions revolve around integration of environmental, social and economic dimensions of development (e.g. expressed in terms of three pillars or a three-legged stool, the triple bottom line (see Elkington, 1997) or a Venn diagram with three intersecting circles). There is usually also some consideration of long-term time horizons with respect to giving consideration to future generations (i.e. inter-generational equity) and the overall environmental, social and economic conditions that they will inherit as a consequence of the decision currently being made. The key point is that different people and institutions have different understandings of the concept and frame sustainability differently (Bond and Morrison-Saunders, 2011).

This, however, requires some clarification of our use of the terms 'sustainable development' and 'sustainability' in this chapter. Some authors consider these to have very different meanings, and Lélé (1991) provided a semantic map which separated notions of sustainability from notions of development in which sustainable development = sustainability + development. Using a triple bottom line approach, some might expect sustainability in terms of economic prosperity, social justice and environmental quality; but we will see that there are others who accept trade-offs between the three. Because of the contested nature of the definitions, we do not risk alienating readers and choose to use sustainability interchangeably with sustainable development. So, achieving sustainability means achieving sustainable development.

As one example (there are many others, see Bond and Morrison-Saunders, 2009) two conflicting framings of sustainability: weak and strong (Pearce *et al.*, 1993; George, 1999; Neumayer, 2010) are important to note which differ with respect to the treatment of natural and human-made capital (see, for example, Cabeza Gutés, 1996). In summary, strong sustainability does not permit the substitution of one of these types of capital for the other, while weak sustainability does as long as the total capital passed onto future generations does not decrease. Thus an example of a new road through pristine forest represents the weak sustainability position where a decline in natural capital is considered acceptable (and hence to be 'sustainable') provided the socio-economic benefits are considered to compensate for the environmental degradation. Most environmentalists would not accept this outcome as being truly sustainable, arguing that all social enterprise is dependent upon a healthy environment and therefore it is not acceptable to continue to erode natural capital. Others (Costanza *et al.*, 1997; Costanza *et al.*, 2007) suggest that considering the ways in which environmental services such as clean air, clean water, energy etc., are delivered into both natural and human systems is a useful way of understanding whether particular environmental relations are sustainable and whether, therefore, particular decisions are justifiable.

We suggest that most institutions, which traditionally have been aligned to the separate environment, social and economic 'silos', are biased at the very least according to whether they advocate strong or weak sustainability (and with further specific interests or biases if supportive of the weak position). Sustainability assessments undertaken in England were found to lead to social and economic benefits relating to the appraised plans (Thérivel *et al.*, 2009) for the sample of 45 examined, but negative environmental effects (this is considered in more detail in Chapter 9). They did not find there to be explicit application of weak sustainability, rather this appeared to arise implicitly as a product of institutional bias.

There are many other contested aspects of sustainability that lead to different framings of the concept. Particular framings favour particular discourses and marginalise others. This potentially leaves sustainability assessment as a generic practice open to failure in the eyes of some observers. The key learning point is that a sustainability assessment must clearly establish the meaning of 'sustainability' or 'sustainable development' as it applies to that assessment for the benefit of all stakeholders in the process. This leads us to the second critical debate relating to the context of a sustainability assessment.

Sustainability and context

An assessment process should be context-specific (Bina, 2008), in that it needs to be flexible and adapt to the different dimensions of context (which Bina indicates are: values; cultural; political; and social). To illustrate the point, consider the notion of sustainability in different sectors. For a typical mining project in Australia, for example, a natural resource would be extracted from the earth and converted into another form; in doing so, land of significant value to Aboriginal people might be degraded. From a strong sustainability perspective, this is unsustainable; from a weak sustainability perspective it can be sustainable. However, is it really feasible to adopt a strong sustainability framing for a mining project where the only option (assuming recycling cannot meet demand) is not to supply raw materials which are in demand? In such a context, one decision has already been made – that development of the resource is needed. In many cases, this changes the context against which sustainability is judged because extraction somewhere is a given. However Howitt (2011: 142) argues that 'context matters – the historical, geographical, social and cultural context in which we undertake research fundamentally shapes what we come to know'. This suggests that the interpretation of sustainability is dependent both on the decision-context (of those in power), and on the context defined by time, place and culture (and in the mining example, the land degradation cannot be acceptable to the Aborigines). This reinforces one of the imperatives set out by Gibson in Chapter 1: that sustainability assessment must respect the context. However, 'context' itself is a plural concept.

This may lead to sustainability assessment being perceived as problematic because it may imply that the meaning of 'sustainability' is uncertain and it may not lend the process substance and allow outcomes of different assessments, each of which may claim to represent sustainability assessment practice, to be compared. The key learning point is, like for critical debate 1, that the meaning of sustainability must be established to avoid confusion or conflict among sustainability assessment stakeholders; the difference is that this must take into consideration the context in which development and decision making will occur for that sustainability assessment, so we cannot establish the meaning of sustainability once in one sustainability assessment, and copy it for others (Pope and Morrison-Saunders make this clear in their review of pluralism in practice – Chapter 7).

Contested time horizons and spatial boundaries

Definitions of sustainable development invariably refer to intra- and intergenerational equity with a very specific consideration of equity in present generations and the level of capital passed down to future generations (whether a weak or strong framing of sustainability prevails). However, evidence suggests that the timescales considered in sustainability assessments are intragenerational at best and, often, are constrained by the nature of the lifetime of the plan or project being assessed (Bond and Morrison-Saunders, 2011). The situation is complicated by arguments that intra- and intergenerational equity are, to an extent, mutually exclusive in that protecting natural capital for future generations does so at the expense of today's poor (Barrett and Grizzle, 1999).

Examples of contested time frames are provided by Rich and Brent Stoffle and Annelie Sjölander-Lindqvist in Chapter 4. What these indicate is that, again, different stakeholders will have different views as to the appropriate timescale to consider in any assessment and there is no simple argument for one view being better than another. Assessment over long timescales is difficult, which may be a reason that it is rarely attempted. In the context of predicting impacts, Gee and Stirling (2004) distinguish between *risk* (where impacts and their probabilities are known), *uncertainty* (where impacts are known but their probabilities are not) and *ignorance* (where neither impacts nor their probabilities are known). Over very long timescales, predictions in sustainability assessment are likely to be based on both uncertainty and ignorance. There is little practice on which to draw for such predictions, and certainly no follow-up studies. Just because something is difficult does not mean that it should be ignored. The key learning point here is that the time frames of a sustainability assessment need to be explicitly identified – as with the previous critical debates, clear unambiguous communication is paramount for effective practice.

Similarly, one of the big drivers of changes in environmental governance in the late twentieth century was the challenge of cross-boundary environmental effects (Schrage and Bonvoisin, 2008), in which the jurisdiction responsible for making decisions and receiving benefits from a decision was different to the jurisdiction in which environmental or social costs were being met. The spatial separation of causes and effects, and the development of increasingly sophisticated global institutions and processes for environmental governance, means that the spatial boundaries and scales at which sustainability effects of any decisions are assessed cannot be restricted to a single jurisdiction and the ways in which cross-boundary and cross-scale effects operate need to be considered. Howitt takes up these concerns in more detail in Chapter 5, but the key learning point from this is that while sustainability assessment will always be framed in a particular jurisdictional and institutional setting, that setting itself needs to be considered in terms of its historical and geographical context, and the ways in which boundaries of sustainability effects are considered needs to be well-justified and explicitly discussed.

Holism versus reductionism

Sustainability assessment typically requires the derivation of indicators, or criteria, which can be used as measures of the state of the socio-economic and biophysical environment and are therefore used as the basis for predictions where a change is proposed (Bockstaller and Girardin, 2003; Donnelly *et al.*, 2007). They also provide the basis for comparing alternatives. However, there is a debate over the degree to which a sustainability assessment should be reductionist and the degree to which it should be holistic (Bell and Morse, 2008). Reductionism we define as breaking down complex processes to simple terms or component parts; for example, selecting a few sustainability indicators to represent the sustainability of a whole system. Steinemann (2000: 640) defines a holistic approach as one which facilitates 'moving away from analyses of isolated risks and toward a broader understanding'.

Evidence currently suggests that the emphasis in sustainability assessment is very much on reductionism, but that the degree of reductionism varies a great deal within particular jurisdictions. For example, we previously reported on variance within sustainability assessment practice in England as well as great contrast with practice in Western Australia simply with respect to the number of indicators applied (Bond and Morrison-Saunders, 2011). These sustainability assessments can be criticised by observers for using the wrong indicators, or too few indicators. From a pragmatic point of view, a large number of indicators leads to an unwieldy, time-consuming and expensive sustainability assessment exercise, and there have already been calls in England to reduce the number of indicators used (Institute of Environmental Management and Assessment, 2006). The reality of the application of sustainability assessment is that some indicators will suggest benefits of particular alternatives, and others will suggest negative impacts for the same alternatives; this inevitably leads to trade-offs and changes the focus of decision making from sustainable development to delivery of the 'least worst' outcome. The key learning point here is that the choice of indicators will determine the characteristics of sustainability considered in a sustainability assessment and consequently the types of alternatives considered and selected in decision making. This can lose sight of the bigger picture and those involved in a sustainability assessment need to take a step back from time to time to reflect on the relevance of the indicators and to satisfy themselves that the right things are being measured.

Process versus outcomes

Just as monitoring and follow-up studies provide the ultimate test for the effectiveness of EIA in achieving its environmental protection goals, it is the outcomes and legacy of sustainability assessment into the future that will establish the ultimate sustainability credentials of any decision-making process. The long time-frames for sustainability considerations mean that points of follow-up and verification may not occur for a considerable time period. This underscores the necessity to get the decision right before action occurs.

Because the nature of any assessment is to 'think before you act' or, in more technical terms, to attempt to predict the future consequences of a proposed activity, a lot of effort is invested in ensuring that a sound or effective process is followed. However there is no guarantee that a 'good process' will ensure delivery of the assessment goals; i.e. that it will automatically equate with good or sustainable development once the proposed activity becomes operational and changes to the physical and social environment actually occur.

This critical debate can be explored by considering the role of courts in assessment disputes. One particular issue, considered by Donna Craig and Michael Jeffery in Chapter 6, is that the courts tend to focus on procedural compliance and not outcomes. Courts typically ensure that steps set out in legislation have been adhered to, but do not (usually) have jurisdiction over the implications of decisions made after assessments are complete (discretion to interpret assessments is left to the elected representatives who tend to make the final decisions). This is not to say the courts are not performing a useful function, the issue is more that legislation is established to ensure assessment takes place (in the name of sustainable development) but does not require sustainable development to ensue. The key learning point here is that a sustainability assessment process must be carefully designed both in terms of the processes to be followed as well as the outcomes that are intended to be delivered. Ideally these would be explicitly stated and linked wherever appropriate. Further, there is an overwhelming need to ensure that institutional commitment to monitoring and follow-up in sustainability assessment is matched by a willingness to undo decisions that prove, with hindsight, to be more damaging than anticipated.

Measuring effectiveness

Having outlined some of the key challenges with designing and undertaking sustainability assessment in practice, our second challenge is to consider how the 'effectiveness' of a sustainability assessment process might be measured. Determining whether an assessment process is 'effective' seems to be a shared goal for academics, investors and policy makers alike. There are literally hundreds of academic papers which consider the effectiveness of environmental assessment, but what do these researchers mean when they consider 'effectiveness'?

An early attempt to take a global look at the effectiveness of environmental assessment was coordinated by the International Association for Impact Assessment and culminated in the *International Effectiveness Study* (Sadler, 1996). In this study 'effectiveness' was defined as 'whether something works as intended and meets the purpose(s) for which it was designed' (Sadler, 1996: 37). However, operationalising this concept to break down the measurement of effectiveness into criteria which can be used as the basis for evaluation is more complex, although progress has been made.

Three types of effectiveness are defined by Sadler (1996):

- *Procedural* – which indicates the extent to which the assessment process properly follows established, or legally mandated, procedures.
- *Substantive* – which indicates the extent to which the goals, or objectives, of the assessment process have been met (this might mean a more sustainable outcome).
- *Transactive* – which considers the extent to which the substantive outcomes are delivered efficiently in terms of cost and time.

Another type is introduced by Baker and McLelland (2003):

- *Normative* – the extent to which the assessment facilitates the achievement of the normative goals. Normative goals are those which are derived from a combination of social and individual norms. These is no universal definition of what such norm are (Gibbs, 1965), but they tend to be considered as standards which society expects conformance with (irrespective of whether we do conform). In the context of sustainability assessment, these norms reflect what we expect the sustainability assessment to achieve, and how it achieves it.

When examining the effectiveness of EIA, Cashmore *et al.* (2004) concluded that the research focus to date has been largely procedural (a typical, and worthwhile, example is Wood, 2003), with some rare inclusion of substantive elements (see, for example, Jones *et al.*, 2005). Theophilou *et al.* (2010: 137) focus on what they argue to be the 'least-researched types of effectiveness' which they consider to be substantive and transactive. So there is some agreement that our understanding of effectiveness is somewhat limited to a narrow set of easily measured parameters dealing with legal procedures. Cashmore *et al.* (2004) make it clear, however, that substantive outcomes of assessment processes are important, but are significantly complicated by the plurality of views on what the purpose of the assessment might be. That purpose needs to be agreed before we can expect different stakeholders to agree that sustainability assessment has been effective. The progress which has been made in the past by studies like the International Effectiveness Study (focussing on EIA) has been based on a defined purpose for assessment (in the 1996 study on EIA this was: to facilitate sound decisions in which environmental consideration are included; and to support the goals of environmental protection and sustainable development).

So we can categorise types of effectiveness and conclude that we know little about how well assessment processes perform except procedurally. A special issue of the journal *Impact Assessment and Project Appraisal* examined further the consideration of effectiveness in impact assessment and concluded that, at present, the 'notion of effectiveness as some sort of absolute measure is untenable' (Cashmore *et al.*, 2009: 93). This suggests that dividing the evaluation of effectiveness into categories (procedural, substantive, transactive and normative) and developing criteria under each of these categories to test against a particular example of sustainability assessment will not tell us whether it is effective. Why? Because a key consideration in evaluating substantive and normative effectiveness is the pluralism (by which we simply mean that there is more than one view) associated with discourses related to the purposes of sustainability assessment. We now explore the relationship between pluralism and effectiveness in more detail (the third challenge).

Pluralism and effectiveness

Pluralism in the context of environmental decision making is discussed by Adger *et al.* (2003). They emphasise that different actors will have different values regarding desirable decision outcomes (for example whether a solution should be sought that favours one species over another), but they also have values based on different ethical premises (for example, whether benefits should be maximised in total, or whether the equitable distribution of benefits should be the goal). In the context of sustainability assessment, the message is that pluralism complicates our consideration of effectiveness because there are different views both on what effectiveness means, and the underpinning arguments for the different views.

Arguably those in charge of setting the objectives for a sustainability assessment are determining the context for effectiveness (Elling, 2009) and, therefore, the developer and the authorities are likely to control the process. Evidence that such control is exerted in the assessment process was found by Cashmore *et al.* (2008) through an in-depth study of environmental assessment cases in the UK, where powerful stakeholders controlled the process, including the selection of other stakeholders to become involved, and the methods to be used. In the context of pluralism, this means that, of all the discourses that might exist regarding sustainability assessment and its purposes, it is the actors who wield power that get to impose their own discourse, both through setting the goals of the assessment and through control over those who are engaged such that only those sharing the same discourse were included. This brings our discussion back to questions of environmental governance and the institutional arrangements governing decision making, accountability and assessment. The need for assessment to operate in the context of power was emphasised by Richardson (2005). So different discourses over the purposes of sustainability assessment exist, but typical practice favours some of those discourses at the expense of others. To put it bluntly, the views of some people are actively ignored which manipulates the assessment.

Hommes *et al.* (2009) cite a case where they consider 'superfluous knowledge' to have been generated in an EIA because it did not address the perceptions of all stakeholders in decision making. They also indicate, in a review of the assessment process associated with a proposed extension of Rotterdam port, that decision makers could use knowledge in a strategic way which was at odds with the views of specific stakeholders. In this particular example, knowledge on ecological impacts was derived from modelling studies and from expert knowledge; however, some modelling work was completed after experts gave their initial views. The experts then wanted the opportunity to re-appraise the impacts, but the decision makers chose to re-interpret their original findings themselves in the light of the modelling results. This reinforces the point made about power above, but here the focus is on knowledge generated in the assessment which will subsequently inform decisions.

Elling (2009: 130) insists that assessment processes should aim for outcomes of mutual understanding in terms of 'truth, rightness and truthfulness'. O'Faircheallaigh (2009) argues, in the context of decision processes which have historically excluded Aboriginal peoples in Australia, that control of the assessment process by indigenous people is an essential prerequisite of effectiveness. Van Buuren and Nooteboom (2009) argue that effectiveness relies on collaborative dialogue to achieve consensus and frame reflection. Owens *et al.* (2004: 1943) argue that assessment processes should 'provide space for dialogue and learning in the making of policies'. All of these authors, although taking different perspectives and working with different assessment tools, have reached a common conclusion that pluralism has to be accommodated by the assessment process.

Bond *et al.* (2011) emphasise that each sustainability assessment process should be undertaken with an explicit aim of 'learning while doing' because this engenders an approach whereby discourses are not assumed, including those surrounding what the goals of the sustainability assessment might be. Cashmore *et al.* (2010: 371) concur with this approach in concluding that 'learning derived from analysing the meaning and implications of plural interpretations of effectiveness represents the most constructive strategy for advancing impact assessment and policy integration theory'. Thus, we should not assume there is a correct model for sustainability assessment, rather, we should consider practice to be a learning experience in terms of the accommodation of different views and the generation and legitimate use of relevant knowledge. For this reason, Jenny Pope and Angus Morrison-Saunders explore current practice related to the accommodation of pluralism in Chapter 7.

Conclusions

The five critical debates examined in this chapter demonstrate how values and circumstances can influence the way observers view the effectiveness of the sustainability assessment process. We strongly advocate that each of these five debates, concerning sustainability definition and context, time horizons and spatial boundaries, being holistic and whether process or outcome based, must be clearly articulated to stakeholders from the design stages onwards in a sustainability assessment process. Failure to communicate these effectively will create uncertainty that will undermine the credibility of the sustainability assessment.

We have also examined the term 'effectiveness' and have begun to understand the different ways of conceptualising it. Underpinning all these has been pluralism and the need to accommodate it and learn from practice. So what scope is there to move forward and develop sustainability assessment such that it can satisfy the effectiveness discourse of a wide variety of stakeholders?

Pluralism underpins all the challenges. The five critical debates illustrate how pluralism leads to contested views which inevitably affect interpretations of effectiveness. The challenge of measuring effectiveness makes it clear that particular criteria for sustainability assessment can be selected and examined in an objective way, but the derivation of the criteria can emphasise particular viewpoints in relation to the five critical debates, and therefore bias the evaluation that is undertaken. In Chapter 8, we build on these challenges to develop a means of evaluation that can accommodate the pluralism.

Notes

1 Following Svarstad *et al.* (2008) we define a 'discourse' as a system of knowledge or beliefs – or a worldview in laypersons' terms

References

Adger WN, Brown K, Fairbrass J, Jordan A, Paavola J, Rosendo S, Seyfang G (2003) 'Governance for sustainability: towards a "thick" analysis of environmental decisionmaking', *Environment and Planning A*, 35, 1095–1110.

Baker DC and McLelland JN (2003) 'Evaluating the effectiveness of British Columbia's environmental assessment process for first nations' participation in mining development', *Environmental Impact Assessment Review*, 23(5), 581–603.

Barrett CB and Grizzle RE (1999) 'A holistic approach to sustainability based on pluralistic stewardship', *Environmental Ethics*, 21, 23–42.

Bell S and Morse S (2008) *Sustainability Indicators: Measuring the Immeasurable?* (London, Sterling, VA: Earthscan).

Bina O (2008) 'Context and systems: thinking more broadly about effectiveness in strategic environmental assessment in China', *Environmental Management*, 42(4), 717–733.

Bockstaller C and Girardin P (2003) 'How to validate environmental indicators', *Agricultural Systems*, 76(2), 639–653.

Bond A, Dockerty T, Lovett A, Riche AB, Haughton AJ, Bohan DA, Sage RB, Shield IF, Finch JW, Turner MM, Karp A (2011) 'Learning how to deal with values, frames and governance in sustainability appraisal', *Regional Studies*, 45(8), 1157–1170.

Bond AJ and Morrison-Saunders A (2009) 'Sustainability appraisal: jack of all trades, master of none?', *Impact Assessment and Project Appraisal*, 27(4), 321–329.

Bond AJ and Morrison-Saunders A (2011) 'Re-evaluating sustainability assessment: aligning the vision and the practice', *Environmental Impact Assessment Review*, 31(1), 1–7.

Cabeza Gutés M (1996) 'The concept of weak sustainability', *Ecological Economics*, 17(3), 147–156.

Cashmore M, Bond A, Cobb D (2008) 'The role and functioning of environmental assessment: theoretical reflections upon an empirical investigation of causation', *Journal of Environmental Management*, 88(4), 1233–1248.

Cashmore M, Bond A, Sadler B (2009) 'Introduction: the effectiveness of impact assessment instruments', *Impact Assessment and Project Appraisal*, 27(2), 91–93.

Cashmore M, Gwilliam R, Morgan R, Cobb D, Bond A (2004), 'The interminable issue of effectiveness: substantive purposes, outcomes and research challenges in the advancement of environmental impact assessment theory', *Impact Assessment and Project Appraisal*, 22(4), 295–310.

Cashmore M, Richardson T, Hilding-Ryedvik T, Emmelin L (2010) 'Evaluating the effectiveness of impact assessment instruments: theorising the nature and implications of their political constitution', *Environmental Impact Assessment Review*, 30(6), 371–379.

Costanza R, D'Arge R, De Groot R, Farber S, Grasso M, Hannon B, Limburg K, Naeem S, O'Neill RV, Paruelo J, Raskin RG, Sutton P, Van Den Belt M (1997) 'The value of the world's ecosystem services and natural capital', *Nature*, 387(6630), 253–260.

Costanza R, Fisher B, Mulder K, Liu S, Christopher T (2007) 'Biodiversity and ecosystem services: a multi-scale empirical study of the relationship between species richness and net primary production', *Ecological Economics*, 61(2–3), 478–491.

Donnelly A, Jones M, O'Mahony T, Byrne G (2007) 'Selecting environmental indicators for use in strategic environmental assessment', *Environmental Impact Assessment Review*, 27(2), 161–175.

Elkington, J (1997) *Cannibals With Forks: The Triple Bottom Line of 21st Century Business* (Oxford: Capstone).

Elling B (2009) 'Rationality and effectiveness: does EIA/SEA treat them as synonyms?', *Impact Assessment and Project Appraisal*, 27(2), 121–131.

Fuller K (1999) 'Quality and quality control in environmental impact assessment'. In: Petts J (ed.) *Handbook of Environmental Impact Assessment. Volume 2 – Environmental Impact Assessment in Practice: Impact and Limitations* (Oxford: Blackwell Science), pp.55–82.

Gee D and Stirling A (2004) 'Late lessons from early warnings: improving science and governance under uncertainty and ignorance'. In: Martuzzi M and Tickner JA (eds), *The Precautionary Principle: Protecting Public Health, the Environment and the Future of Our Children* (Copenhagen: WHO Regional Office for Europe), pp.93–120.

George C (1999) 'Testing for sustainable development through environmental assessment', *Environmental Impact Assessment Review*, 19(2), 175–200.

Gibbs JP (1965) 'Norms: the problem of definition and classification', *The American Journal of Sociology*, 70(5), 586–594.

Grey W (1993) 'Anthropocentrism and deep ecology', *Australasian Journal of Philosophy*, 71(4), 463–475.

Hacking T and Guthrie P (2008) 'A framework for clarifying the meaning of triple bottom-line, integrated, and sustainability assessment', *Environmental Impact Assessment Review*, 28(2–3), 73–89.

Hommes S, Hulscher SJMH, Mulder JPM, Otter HS, Bressers HTA (2009) 'Role of perceptions and knowledge in the impact assessment for the extension of Mainport Rotterdam', *Marine Policy*, 33(1), 146–155.

Howitt R (2011) 'Chapter 8: knowing/doing'. In: Vincent J, Del Casino J, Thomas ME, Cloke P, Panelli R (eds), *A Companion to Social Geography* (Chichester: Wiley-Blackwell) 131–145.

Institute of Environmental Management and Assessment (2006) 'SEA forum report'. [Online] Available at: <http://www.iema.net/stream.php/download/reading room/article/2006%20SEA%20Report.pdf> (accessed 5 March 2009).

Jacob M (1994) 'Sustainable development and deep ecology: an analysis of competing traditions', *Environmental Management*, 18(4), 477–488.
Jones CE, Baker M, Carter J, Jay S, Short M, Wood C (eds) (2005) *Strategic Environmental Assessment and Land Use Planning: an International Evaluation*. (London: Earthscan Publications).
Lélé SM (1991) 'Sustainable development: a critical review', *World Development*, 19(6), 607–621.
Neumayer E (2010) *Weak Versus Strong Sustainability: Exploring the Limits of Two Opposing Paradigms* (Cheltenham: Edward Elgar).
O'Faircheallaigh C (2009) 'Effectiveness in social impact assessment: Aboriginal peoples and resource development in Australia', *Impact Assessment and Project Appraisal*, 27(2), 95–110.
Owens S, Rayner T, Bina O (2004) 'New agendas for appraisal: reflections on theory, practice, and research', *Environment and Planning A*, 36(11), 1943–1959.
Pearce D, Turner RK, O'Riordan T, Adger N, Atkinson G, Brisson I, Brown K, Dubourg R, Fankhauser S, Jordan A, Maddison D, Moran D, Powell JC (1993) *Blueprint 3: Measuring Sustainable Development* (London: Earthscan Publications).
Richardson T (2005) 'Environmental assessment and planning theory: four short stories about power, multiple rationality, and ethics', *Environmental Impact Assessment Review*, 25(4), 341–365.
Sadler B (1996) *International Study of the Effectiveness of Environmental Assessment Final Report – Environmental Assessment in a Changing World: Evaluating Practice to Improve Performance* (Ottawa: Minister of Supply and Services Canada).
Schrage W and Bonvoisin N (2008) 'Transboundary impact assessment: frameworks, experiences and challenges', *Impact Assessment and Project Appraisal*, 26(4), 23–38.
Steinemann A (2000) 'Rethinking human health impact assessment', *Environmental Impact Assessment Review*, 20, 627–645.
Svarstad H, Petersen LK, Rothman D, Siepel H, Wätzold F (2008) 'Discursive biases of the environmental research framework DPSIR', *Land Use Policy*, 25(1), 116–125.
Theophilou V, Bond A, Cashmore M (2010) 'Application of the SEA Directive to EU structural funds: perspectives on effectiveness', *Environmental Impact Assessment Review*, 30(2), 13–44.
Thérivel R, Christian G, Craig C, Grinham R, Mackins D, Smith J, Sneller T, Turner R, Walker D, Yamane M (2009) 'Sustainability-focused impact assessment: English experiences', *Impact Assessment and Project Appraisal*, 27(2), 15–68.
Van Buuren A and Nooteboom SG (2009) 'Evaluating strategic environmental assessment in The Netherlands: content, process and procedure as indissoluble criteria for effectiveness', *Impact Assessment and Project Appraisal*, 27(2), 14–54.
Williams CC and Millington AC (2004) 'The diverse and contested meanings of sustainable development', *The Geographical Journal*, 170(2), 99–104.
Wood C (2003) *Environmental Impact Assessment: A Comparative Review* (Edinburgh: Prentice Hall).
World Commission on Environment and Development (1987) *Our Common Future* (Oxford: Oxford University Press).

4 Contested time horizons

Richard W. Stoffle, University of Arizona, Tucson, USA
Brent W. Stoffle, National Oceanic and Atmospheric Administration, Miami, USA
Annelie Sjölander-Lindqvist, University of Gothenburg, Gothenburg, Sweden

Note: Ideas expressed within this chapter are solely those of the authors and do not necessarily reflect any views, opinions or positions of NOAA.

Introduction

This chapter explores the implications of contested time horizons for sustainability assessment. Sustainability is normally defined as when a human/natural system's qualities (fundamental to system function) remain the same or increase over time (Bell and Morse 2010, p.12). The timescale over which sustainability is considered and monitored is both a part of its definition and a variable by which it is evaluated (Bell and Morse 2010, p.15). Most timescale discussions focus on the issue of equity for future generations (Dresner 2002, p.2) but even the measure of a generation and how many generations should be considered are open to debate (Bond and Morrison-Saunders 2011, p.4). Sustainability over time is often discussed in terms of simple phrases such as "don't cheat your kids" but *time* as a dimension of sustainability is much more complex.

Proposals to build on, modify, or preserve places occur in different social, cultural, and political settings thus producing contested approaches to what constitutes appropriate or acceptable impact assessment time horizons. The arguments for and against the use of longer or shorter time horizons can be based on science, epistemological premises, values, and value priorities. Science is used to argue for one position and ethics can be used to argue for another. When people of different cultures are involved in the debate, fundamental beliefs about the nature of the world add almost irresolvable epistemological differences. Often economic values (cost overruns or opportunity costs) are used as an *Occam's razor* (i.e. a succinct way to resolve debates). Sometimes a lack of longitudinal scientific observations is used to argue for shortening the projected impact assessment time frames so as to fit the available data.

The notion of time horizons is similar to a *project footprint*; each has a temporal and spatial focus that locates the outer boundaries of the assessment task. As such, these are critical variables because they indicate which resources and over what time frame will be taken into consideration. Retrospective impact studies document the need to have extended the time horizon of assessments. Outstanding among these are the actual impacts of the Columbia River dams, which were constructed in the 1920s and 1930s in the Pacific Northwest of the United States. The adverse environmental impacts of those projects, highlighted by the drastic reduction and extinction of fish species, especially salmon, have called into question the design, placement, operation, and even the construction of dams (Ortolano and May, 2004). Hindsight affords recognition that many projects have had similarly unimaginable catastrophic consequences and should have either never been approved or waited until the science of human and natural systems caught up with the technology to modify them.

Adam (1998) uses social philosophy to analyze how and why these time and space assessment failures occur. She maintains that when assessments have minimal scope and variables this occurs because of strategic decisions by project proponents. Her analysis uses a concept called *timescapes* (Adam 2006, pp.143–145), which folds into an integrated sustainability assessment analysis the motives of scientists, project proponents, and government regulators when they chose to define minimal timescapes. Adam maintains that regulators, scientists, and proponents choose minimal timescapes in order to appear to be in control of time and space impacts, and thus assure others (especially the public and stakeholders) of their ability to predict and mitigate impacts. Instead of actually being able to better predict and control, however, Adam maintains that they actually mask their inability to do either. Timescapes are minimized in order to avoid grappling with outcomes that are increasingly unknown and potentially difficult to mitigate. This apparent increase in mastery of sustainability assessment variables and outcomes due to minimal timescapes is a rhetoric that has accompanied the emergence and spread of modern technological innovations like nuclear power and bioengineering (Adam 2006, p.146).

Building on Adam's analysis and the retrospective assessments of others, it can be argued that if a sustainability assessment study utilizes the longest perceivable time horizons and the widest project footprint it will necessarily report on the nature of uncertainty and exposure to risks generated by a project and subsequently require these to be addressed in the project assessment and approval processes. Impact concerns raised by a sustainability assessment, however, tend to reduce the probability that the proposed project will be approved for development. Thus it is in the interest of project proponents to diminish concerns and emphasize potential benefits (Corvellec and Boholm, 2008). As a general rule of thumb, the further you predict impacts into the future the longer you should know about them in the past.

Time horizon issues

Time horizon issues generally derive from sustainability assessment pressures to consider *intergenerational impacts and equity* (Gibson *et al.*, 2005, pp.103–105). Many project assessments have a cost-to-benefit analysis worked out for less than a few decades. A sustainability assessment, however, should consider distant future impacts to people and nature. A sustainability assessment methodology argues for using time horizons that range from more than a couple of generations to the extreme 100,000 year horizon of nuclear radioactive waste isolation (Bond and Morrison-Saunders, 2011).

Considered here are four key time horizon issues: (1) the length of time of analysis, (2) the culturally defined value of the resource, (3) the resilience of the resource, and (4) whose problem is being solved by the development – a local one or a national one.

Length of time horizons

Each assessment process establishes a time horizon for all relevant impact studies. Often this time horizon is as short as the project proponent and the regulatory agency can achieve. From their perspective the shortest arguable time horizon is best because the resulting predictions have the highest *confidence levels*. In general, the confidence levels of predictions diminish with longer time horizons. Scientists simply do not know what might occur at certain time horizons because deep time baselines and long-term monitoring data are not available. Some development critics also argue that shorter time horizons are chosen in order to avoid raising difficult issues that may slow down, make more expensive, or even eliminate a proposed project.

Cultural value of resource

It is important to know about the cultural value of one or more resources being impacted by a proposed project. The term *cultural centrality* is used to emphasize not only the absolute cultural value of a resource but its role in sustaining other aspects of society and its place in local, regional, and world ecology. Resources become culturally central over generations during which time many have proven their utility in supporting and integrating the society. Thus, these resources demand longer-impact time horizons because of the cultural cost of modifying, losing, or replacing such resources. Damage to culturally central resources can rarely be mitigated, thus they often constitute go-no-go variables in the assessment of project impacts.

Resilience of resource

Resilience is a term that has emerged in common public use as well as in the biological and social sciences. Interestingly its popularity began just as society lost confidence in its ability to persevere in the face of new risks. Perhaps the two are related. As natural and social disasters increase in frequency and intensity, the issue of lifeway survival becomes increasingly salient. Resilience is used in this analysis at the societal scale of size and complexity, not at the individual or small group scales. Resilience is about a social condition that occurs when people in a stable society become traditional, learn about their ecosystems, and adjust their adaptive strategies to protect them from natural and social perturbations (Stoffle *et al.*, 2003). The natural and cultural resources that help produce this social condition are culturally central to the society. Resilience is concerned with the magnitude, frequency, and kinds of disturbance that can be absorbed without the society and its natural system undergoing fundamental changes.

Who absorbs costs or profits from project impacts?

All projects have *costs and benefits*. A project like a Marine Protected Area designed to preserve or restore the ocean can have adverse impacts such as shifting nature/human relationships and reducing resilience to both nature (Castilla, 1993) and people (Stoffle and Minnis, 2008). Although economic evaluations of costs and benefits are common, certainly some results cannot be measured in money (Rosa and Short, 2004). There is thus the issue of how to measure and even whether or not to measure impacts that must be discussed qualitatively.

Who is absorbing project costs and who is receiving benefits is another key issue. Generally impacts are studied as being absorbed across various spatial scales by individuals, communities, regions, nations, and internationally. Project benefits tend to be front-loaded both by proponents and in assessment procedures, while significant costs that occur much later, such as poorly managed wastes and decommissioning impacts, are discounted or overlooked. Large projects tend to benefit regions and powerful businesses and their costs tend to be absorbed by local communities. The short-term benefits in the form of employment, economic activity, tax revenues, and various positive multipliers overshadow likely longer-term negative consequences. Often there is an assumption that technological or institutional solutions to long-term risks can be relied on to reduce the costs in the future. Nuclear power plants, for example, were argued for because they were environmentally cleaner than coal, but in retrospect it is now clear that enriched uranium fuel rods, as the end products of uranium mining and processing, create a massively long time horizon pollution legacy (Drottz-Sjöberg, 2010; Stoeglehner *et al.*, 2005).

Cases

The following cases illustrate these four time horizon issues in terms of a specific project that has impacted a local community and its social, cultural, and ecological context. The cases are (1) a Southern Paiute Solar Calendar in southern Utah impacted by a high voltage power line and subsequent energy corridor, (2) a rum distillery impacting the ability of local fishers to maintain their way of life in St Croix, U.S. Virgin Islands, and (3) a Swedish railway tunnel cut into a water-filled hill polluting local villagers and threatening their way of life. The cases document the weakness of the short time frames that were used in these Environmental Impact Assessments (EIAs) and thus they argue that the longer time horizons required in sustainability assessment studies would have identified and potentially avoided many of these adverse impacts.

Case #1: Solar Calendar in southern Utah

This case is focused on impacts from a large electrical power line, the Intermountain Power Project (IPP) and its subsequent impacts over a 25-year period on a traditional Solar Calendar and pilgrimage trail in the western US (Stoffle *et al.*, 2008a).

Setting

Solar Calendars are culturally central in the lives of Southern Paiutes. This calendar is located in a very small cave high up on the face of a tall red vertical sandstone outcrop. The place is dramatic with the brick red sandstone standing up from a generally flat and grey sedimentary bench (see Figure 4.1). The bench is covered with trees and a dozen or more types of Indian medicine plants. The pink sandy ground which is produced from eroding sandstone is covered with small chips of brightly colored stone. The climb onto the vertical sandstone is extremely steep and thus has a sharp drop-off to the ground below. The view from the cave is important because it looks out on the surrounding landscape, the path leading up to the cave, and the sacred plants surrounding the path. Once on a high elevated shelf the cave can be climbed into. Immediately in the cave is a large flat polished stone which seems to almost fill the cave. The cave and its clean pink sand floor contrast sharply with the stone which has been totally smoothed to prepare a massive and apparently integrated set of peckings. At the entrance portion of the stone, grooves have been cut into the outer portion of the stone. The top of the cave has a v-shaped break just above the grooves. Light falls on the cuts at various times making the stone a time marker which, with the ceremonial cycles displayed in the peckings, give the cave its calendar functions.

The decision to permit construction of the high-voltage power line opened up motor vehicle access in a formerly isolated piñon forest. After this decision, the power line corridor and its associated construction and access road became a *sacrifice zone* (a term for an area which is targeted for future development because it has been impacted once) in which five utility projects would be permitted. A pilgrimage trail and a sacred Solar Calendar located in this Utah

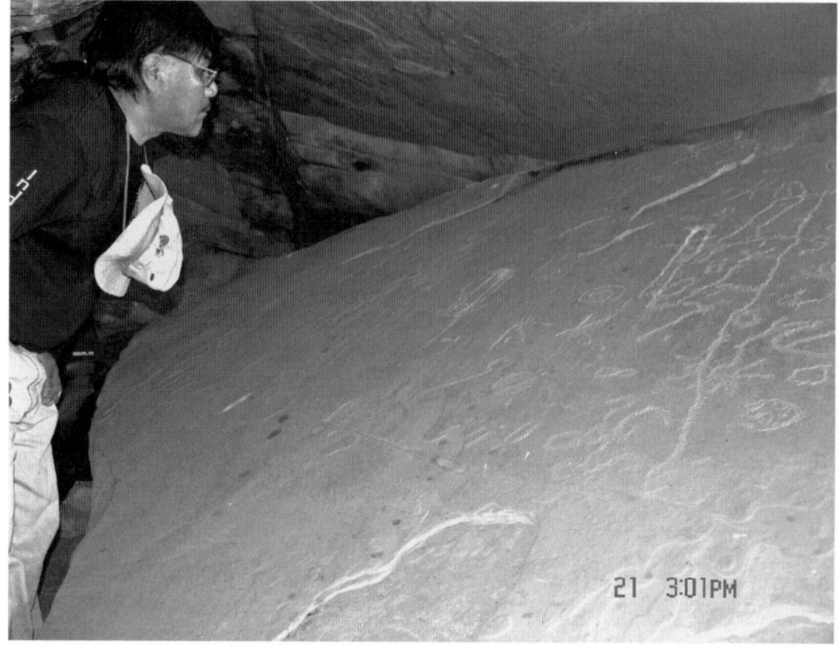

Figure 4.1a

Solar calendar illustrating peckings

forest were damaged because they were not considered in the IPP and subsequent impact assessments.

The case is also about the problems of predicting the future (see Luhmann's chapter on the "Future at Risk", 2006), especially without access to *analog cases* which can be used to better predict the outcomes of environmental decisions. This is now an analog case where we can look back and see where assumptions were wrong or right. This case is bracketed by two large-scale EIAs. It is also a special case because it analyzes *cumulative impact trajectories* which are rarely anticipated when a project opens access to remote areas.

Project EIA studies

The IPP EIA involved studying potential impacts of one 500KV electrical power line and its associated access roads along more than 500 miles of study corridor ranging from Delta, Utah to a power station in the central Mojave Desert in California (Stoffle and Dobyns, 1983). In 1983 (Stoffle *et al.*, 1983), Indian people visited a portion of the IPP study corridor that passed near the Solar Calendar. When they observed the proximity of the IPP proposed corridor they insisted on being interviewed and breaking a previous tribal tradition of not telling non-Indians about the Solar Calendar. In 2006, another EIA study

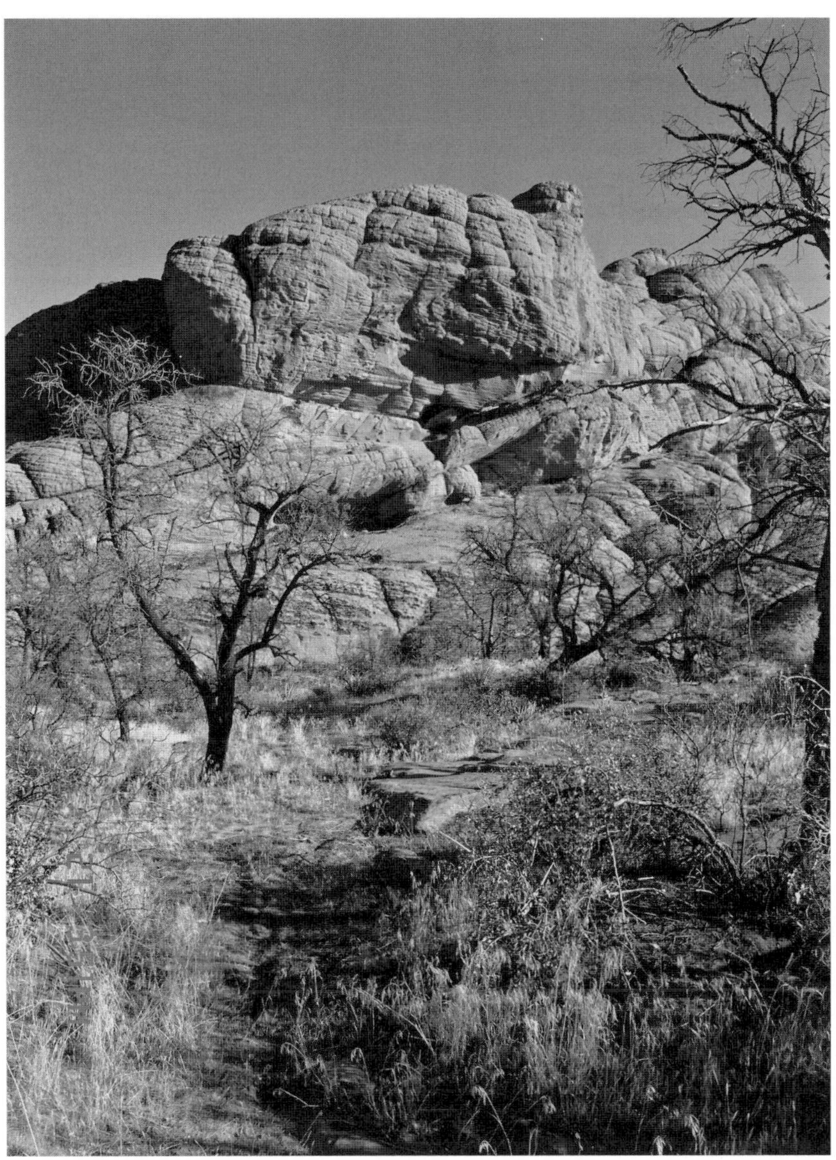

Figure 4.1b

The isolated sandstone in which the solar calendar is located

related to a National Park Service's proposed management plan for the Old Spanish Trail (OST) permitted further interviews at the Solar Calendar (Stoffle *et al.*, 2008b). The OST interviews documented 25 years of cumulative impacts to the site (Stoffle *et al.*, 2008a).

Time horizon issues

Assessing the cumulative impacts of unique proposals that involve building in isolated areas and potentially changing the relationships of traditional people to the land will always be difficult and may at some level be impossible to predict. The Solar Calendar in question is a culturally central and thus a unique power place that has been used to tell time and to regulate ceremonial activities for thousands of years. The 1983 IPP EIA predicted no adverse impacts due in part to the isolation of the area and a very narrow definition of the footprint of the power line. The Indian tribes involved disagreed with that assessment. The decision put at risk the long term ceremonial needs of Indian people for the immediate electrical needs of the people of Southern California.

After decades of assessing utility corridors, it is now clear that they do increase access to natural areas and consequently expose them to further development, increased use, and damage. These findings support Adam's assessment (1998) that when policy makers use short-term cost–benefit analysis they are apt to overlook larger and longer timescapes. The findings also support Cebulla's analysis (2007) which demonstrates that social risks are not equally absorbed by people from different social classes. Tollefson and Wipond (1998) point out that aboriginal people can have different spatial and temporal sustainability assessment issues than the broader society around them. Here Southern Paiute people are shown to have absorbed damage to a unique pilgrimage trail and Solar Calendar so that the people of the nation state can reduce the collective costs of energy by taking transmission shortcuts through isolated areas.

Case #2: Reefs, marine fishermen lifeways in St Croix, U.S. Virgin Islands

This case is about the long-term social and natural impacts that have failed to be addressed in a series of EIA-type decisions over 100 years in St Croix, United States Virgin Islands (USVI). The case focuses on the consequences of effluent from rum production on ecological and social relations. Clearly here the short time frame interests of a large corporation have outweighed the multigenerational livelihood interests of traditional fishing communities.

Setting

Throughout the Caribbean, fishing has retained an important role in people's lives. Indigenous populations utilized marine resources for sustenance and trade. The following centuries saw waves of European settlers and unfree laborers entering the region. Throughout settlement, slavery and post-emancipation times, local communities used marine resources as a means of providing food and garnering an income (Stoffle and Stoffle, 2007). Fishing and marine resources have always been an important component of the local culture of St Croix, which is officially considered a *fishery dependent community* (Stoffle *et al.*, 2009).

There are numerous human and natural perturbations that threaten St Croix's fishing. Some of these can be controlled, including a decision to cut mangroves for new coastal development or the reduction or elimination of highly destructive fishing gear, and others cannot be, such as hurricanes or earthquakes. One perturbation relates to effluent spewed from the Cruzan Rum factory. Studies from the 1940s (Brown, 1944) demonstrated the toxicity of effluent from alcohol production (Grismer *et al.*, 1999; Sen and Bhaskaran, 1962) and rum effluent in the ocean (Fuentes *et al.*, 1983). A study of five rum factories concluded that "rum effluent discharge is harmful to the marine environment resulting in fish kills and reductions in marine microorganisms such as planktonic and heterotrophic organisms" (American University, n.d.). These studies should have been an essential component of any impact assessment and decisions regarding regulation of effluent.

So constant is the discharge that it is nicknamed "the brown reef." When flying onto or off the island shown in Figure 4.2 a brown streak that looks to be a reef can be seen just offshore of the southwest coast. A *Virgin Island News* article documents the impact of effluent released from the Cruzan Rum factory. This article supported many of the local fishermen's reasons why they prefer not to fish in the area, even though there are sometimes mitigating factors that force them to fish there. The article states that:

> (W)hen EPA Administrator Bill Muszynski flew over the south shore of St. Croix, he couldn't help but see the long brown streak of effluent from the Cruzan Rum factory that stains the turquoise water. He said the rum factory had a Clean Water Act waiver from the EPA to discharge its molasses byproduct since 1992. Based on earlier studies, the discharge was deemed unsightly but not harmful to the marine environment. Now however, the EPA suspects it might be toxic and is looking into imposing new restrictions.
>
> (Arawak Roots, n.d.)

Time horizon issues

Cruzan Rum has been produced on the island of St Croix for over 200 years by the Nelthropp family. Throughout this period effluent has been pumped

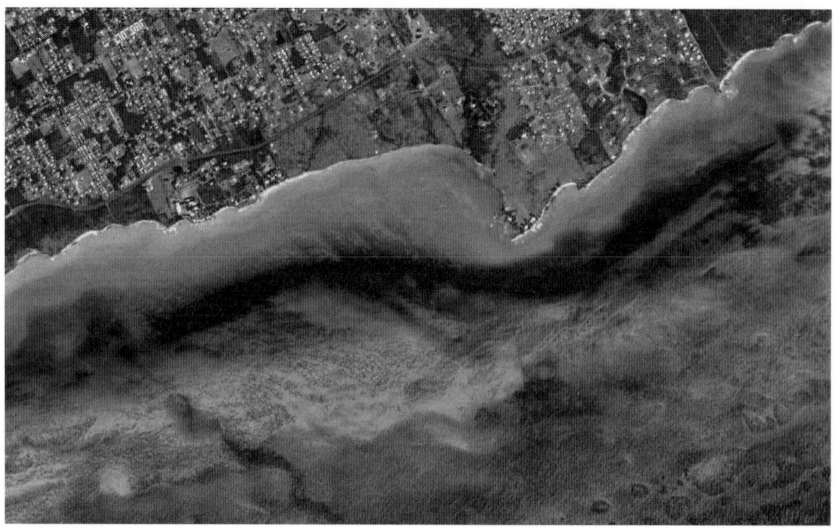

Figure 4.2

Aerial view of effluent on the south shore of St Croix
Source: Google Earth

into the sea. Initially the effluent was deemed to be harmless to the environment; later people realized that the effluent is toxic. Neither the U.S. nor the V.I. government stopped the effluent from being pumped into the ocean nor forced the plant to meet EPA standards. Since 1992, Cruzan Rum has had a waiver which allows the pumping of effluent into the sea.

Cruzan fishermen and local residents have long protested the discharge, which clearly impacted the fish in terms of taste and abundance. In 2005, while B. Stoffle was interviewing local fishermen at the La Reine fish market, a woman arrived with a half-cooked parrotfish in a pan. She said that she could tell where it was from because of the way it smelled. She demanded a refund and refused to accept new fish from the same fisherman. She cursed him for trying to "poison" her and her family. Because one's reputation is so critical in small island society, there is a great fear of negative gossip.

It is necessary to place effluent pollution within the context of which marine areas are available for bottom fishing. Bottom fish or reef fish are the primary food fish for local residents. These are species that locals from all class and ethnic backgrounds consume at relatively inexpensive prices. In addition, high-valued species such as conch and lobster live in and pass through these areas. In St Croix the bottom fishing areas have diminished due to (1) expansion of Buck Island National Park, (2) creation of East End Marine Park, (3) implementation of fishery policy seasonal closings such as the Virgin Snapper spawning grounds, (4) de facto closures due to pollution from the Anguilla Waste Dump and Cruzan Rum effluent, (5) military activities such as submarine movement, and (6) Homeland Security associated with the Hovensa Oil Refinery. The cumulative impacts of diminished fishing areas are to shift fishing pressure to the remaining fishing grounds, which now are unable to sustain new fishing pressures. Fishing as a way of life for the people of St Croix is threatened. The spatial and temporal impacts of effluent pollution must be understood in the context of many other marine issues – a footprint that would have normally been considered in a sustainability assessment study.

Recently the V.I. government entered into a 30-year agreement to bring a new state-of-the-art Captain Morgan rum factory to St Croix. The V.I. legislature agreed to a similar 30-year deal with Cruzan Rum which will build a new and larger factory. Both will be brought up to EPA standards. The new Cruzan Rum plant will help prevent 150,000 lbs of chemical oxygen demand that is now discharged each day into coastal waters. The sustainability of rum production is now required by law.

Case #3: Hallandsås railway tunnel in Sweden

This case deals with cultural differences in the substance and temporal frame of risk perception and who profits and who pays for the improvements to the national railroad system in Sweden (Boholm, 2000).

Setting

The setting for this case (Binde, 2000, pp.54–59; Sjölander-Lindqvist, 2004a) is Båstad (population 14,244) an old idyllic seaside town, dating back to medieval times. It is located on the northern slope of the Hallandsås Ridge (226 meters above sea level at its highest point) in a beautiful hilly rural landscape. Hallandsås ridge holds a large supply of groundwater. The Bjäre Peninsula extends west of Båstad and has a characteristically mild and humid climate, which together with rich soils makes the region very well suited for agriculture. This part of Sweden is renowned for its unseasonably early vegetables, its new potatoes, strawberries, and fruit. It is an old agricultural landscape with many prosperous dairy farms and cattle breeders. Flora and fauna include species that are of national interest and part of the area is classified as a Nature Reserve. Proximity to the sea and the scenic landscape has made the area attractive for summer residents. Båstad has a long tradition as a fashionable resort for vacation and recreation.

The project and its environmental assessment

In 1991 after geological and economic investigations, the Swedish government decided to build an 8.6 kilometer double-tube railway tunnel through the Hallandsås ridge (Figure 4.3). It was argued that a tunnel constituted a necessary means to proffer efficient and modern railway traffic for the transportation of goods and people and to improve safety standards. It was also declared important to extend and upgrade the railway route on the Swedish West Coast in order to promote economic growth in Europe. The National Rail Administration contracted a construction company to build the train tunnel. A specially designed giant drill was used initially, but the machine did not work as planned. Resting on the soft ground, the extremely heavy machine could not be kept stable in the geological conditions of the ridge and had to be abandoned. Drilling work continued with traditional blasting methods. In 1996, a new contract was made and the tunnel project continued with conventional technology and the injection of tightening material to stop the heavy leakage of groundwater. Since the start of the project, considerable amounts of water had been leaking through the tunnel walls and local water resources utilized for a range of activities were affected. Many private wells, dams, and streams dried out. During 1996 it became clear that the conventional sealant for concrete could no longer be used because extremely high water pressure would not let the concrete solidify. In spring of 1997 the problem of the falling groundwater levels became acute. The contractor decided to start sealing the tunnel walls with Rhoca Gil, which is an industrial sealant,

Figure 4.3

Hallandsås Ridge tunnel

produced by the French company Rhône-Poulenc. It contains acrylamide and methyloacrylamide, both known to be carcinogenic and mutagenic. More than 1,400 tonnes of the sealant was injected into the walls of the tunnel, but some of it entered and polluted the groundwater with highly toxic substances known to be carcinogenic and to affect the nervous system.

Within a few months several cows fell severely ill and had to be slaughtered and dead fish were found. A major stream was found to be contaminated and the municipal government declared Hallandsås a risk area. Vegetables, milk, and meat from the area were banned, farmers had to dispose of their products and 370 farm animals were destroyed. Vegetables were left rotting in the fields and in greenhouses. Nine dairy farmers were not allowed to sell their milk and scenes of farmers dumping tonnes of milk into their urine reservoirs were depicted repeatedly in the news media. The annual traditional elk hunt was banned and the Board of Food advised the public not to eat any game or fish from the area. Water wells were found to be polluted and households underwent medical examinations.

Public protests began soon after the announcement of the tunnel pollution and it became a major public event that eroded trust for the National Rail Administration. Delays, pollution mitigations, and lining the tunnel tubes with different sealants have caused massive cost overruns. Clearly unimaginable risks to the local people and the environment were not considered in the initial EIA and thus key issues were not addressed and the best possible drilling and construction practices were not utilized.

Time horizon issues

This tunnel became a major risk example for Sweden. Imagine a huge tunnel being built through a ridge situated in a pastoral agrarian landscape with high nature values. Underground, engineers, construction builders, and officials from the National Rail Administration struggled with construction work that was made increasingly difficult because of complex geology consisting of water-filled crevices and unstable clay, and at places a water pressure exceeding 100 bar. Water leaked into the tunnel and sealing the tunnel walls proved highly problematic. On top of the ridge, farmers who base their livelihood on agriculture and animal husbandry struggled to survive. Traditionally the water makes the soil fertile, the land is suited for cattle, and the artesian water makes wells flow easily. Construction of the tunnel contaminated the environment and the tunnel de-watered the landscape.

A risk perception conflict developed between the local farming community and other affected local residents on the one hand and the tunnel builders, the entrepreneurs, and the Rail Administration officials on the other. Communication between the two social groups revealed disparate socially constructed views of the risks (Boholm, 2009). The Rail Administration identified the ridge and its geological characteristics (such as an extremely high groundwater pressure) as a risk for the tunnel itself, which was framed as a complex, expensive, and vulnerable technical artifact needed by the Swedish people. In contrast, the high groundwater pressure was a positive value at stake from the perspective of the local community members who understood the importance of maintaining it as a prerequisite for the survival of the farming community and for local ecology (Sjölander-Lindqvist, 2004a). The water-rich geology of the ridge posed risks for the tunnel, in terms of technological complications and cost overruns. The local farmers, from their view, identified the tunnel as a risk to the ridge, its sensitive ecosystem, and the future of the entire local social community (Sjölander-Lindqvist, 2004a).

These two social groups share the same environment but they have diverse realities and life worlds, different temporal world views, and identify risks completely differently. The tunnel builders focused on risks associated with their livelihood which was dependent on building the tunnel. They were working for the national good. The farmers focused on risks to their traditional livelihood which include agriculture, animal husbandry, and dairy farming. The farmers thought in terms of furthering the survival of local farming traditions and making it possible to continue their way of life into the distant future (Sjölander-Lindqvist, 2004b).

This case addresses cultural differences in risk perception and the implication of using different timescapes during the assessment of project impacts. Again the issue of who profits and who pays for the improvements to the national railroad systems in Sweden is at issue, and it was not considered in the pre-project planning and still has largely not been addressed by the Swedish government. The tunnel impact assessment weighted the sustainability of Sweden against short-term economics and engineering challenges. A sustainability assessment study would have caused these national benefits to be weighed against long-term adverse impacts to the sustainability of a local Swedish population and environment.

Discussion

Sustainability assessment studies broaden spatial and temporal analysis in the assessment of proposed project impacts. Broader timescapes are difficult to assess but nonetheless cause potentially critical variables to be studied and considered. Projects can destabilize social and natural systems due to impacts that last over generations, thus impacts should be identified and considered in the planning stages to reduce the potential for unimaginable changes to social and natural systems. Each case presented here illustrates that multigenerational adverse impacts can derive from a failure to consider longer timescapes and failure to integrate the concept of sustainability into project siting, operations, and monitoring. These cases are but the tip of a phenomenon that appears to be increasing in modern times.

Accidents and unanticipated impacts will still happen, so people and the environment will continue to experience unimaginable perturbations. Safer decisions, however, are expected to be made by using sustainability assessment studies. The testimony of Nancy Sutley, Chair of the Council on Environmental Quality (Sutley, 2010), before the U.S. Congress regarding the tragic explosion and subsequent release of oil into the Gulf of Mexico by the Deepwater Horizon is instructive. She concludes that a practice of Federal Agencies issuing *Categorical Exclusions* (which exclude certain project types from the need for *ex-ante* assessment) for potentially dangerous projects, like deep-ocean drilling, has been a national mistake and will be changed.

Sustainability assessment is the logical and ethical extension of half a century of environmental impact studies and public education. State of the art does not mean standard of practice as is evidenced by sustainability assessment not being incorporated into environmental laws and regulations in the majority of countries (including the U.S.A. and Sweden). Despite widespread pressure from the environmental assessment community for stronger regulations, these are perceived as threatening weakened economies and unstable political situations. Consideration of longer-impact time horizons is recognized worldwide as helping to secure society despite calling into question many short-term benefit development projects, but these need to be embedded in a properly conducted sustainability assessment, respecting the sustainability imperatives outlined in chapter 1, and encompassing the spatial and cultural issues outlined in the next chapter.

References

Adam B (1998) *Timescapes of Modernity: The Environment & Invisible Hazards* (London: Routledge).

Adam B (2006) *Time* (Cambridge, UK: Polity Press).

American University (no date) *Rum Trade: From Slavery to the Present.* American University. [Online] <http://www1.american.edu/TED/rum.htm> (accessed 1 November 2011).

Arawak Roots (no date) *Current Events in the Caribbean.* Arawak Roots. [Online] <http://www.arawakroots.com/news_1999–2000.php> (accessed 1 November 2011).

Bell, S and Morse, S (2010) *Sustainability Indicators: Measuring the Immeasurable?* (London: Earthscan).

Binde P (2000) "Case Studies of Railway Controversies". In: Boholm Å. (ed.) *National Objectives – Local Objections: Railroad Modernization in Sweden* (Gothenburg: Centre for Public Sector Research, Göteborg University) 39–68.

Boholm Å (ed.) (2000) *National Objectives – Local Objections: Railroad Modernization in Sweden* (Gothenburg: Centre for Public Sector Research, Göteborg University).

Boholm Å (2009) "Speaking of Risk: Matters of Context", *Environmental Communication*, 3(3), 335–354.

Bond A and Morrison-Saunders A (2011) "Re-evaluating Sustainability Assessment: Aligning the Vision and the Practice", *Environmental Impact Assessment Review* 31, 1–7.

Brown EM (1944) "Disposal of Waste from Brandy and Molasses Distilleries from the Viewpoint of the Industry", *Sewage Works Journal* 16(5), 949–951.

Castilla J (1993) "Humans: Capstones Strong Actors in the Past. Present, Coastal Ecological Play". In: McDonnel M and Pickett S. (eds) *Humans as Components of Ecosystems* (New York: Springer-Verlag) 158–162.

Cebulla A (2007) "Class or Individual? A Test of the Nature of Risk Perceptions and the Individualisation Thesis of Risk Society Theory", *Journal of Risk Research*, 10(2), 129–148.

Corvellec H and Boholm Å (2008) "The Risk/No-risk Rhetoric of Environmental Impact Assessments (EIA): The Case of Offshore Wind Farms in Sweden", *Local Environment*, 13(7), 627–640.

Dresner S (2002) *The Principles of Sustainability* (London: Earthscan).

Drottz-Sjöberg B-M (2010) "Perceptions of Nuclear Waste Across Extreme Time Perspectives", *Risk, Hazards & Crisis in Public Policy*, 1(4), 231–253.

Fuentes F, Biamon E, Hazen T (1983) "Bacterial Chemotaxis to Effluent from a Rum Distillery in Tropical Near-shore Coastal Water", *Applied and Environmental Microbiology*, 46(6), 1438–1441.

Gibson RB, Hassan S, Holtz S, Tansey J, Whitelaw G (2005) *Sustainability Assessment: Criteria and Processes* (London: Earthscan).

Grismer M, Carr MA, Shepherd HL (1999) "Fermentation Industry", *Water Environment Research* 71(5) 805–812.

Luhmann N (2006) "The Future at Risk". In: Luhmann, N, *Risk: A Sociological Theory* (Berlin: de Gruyter) 33–50.

Ortolano L and May C (2004) "Appraising Effects of Mitigation Measures: The Grand Coulee Dam's Impacts on Fisheries". In: Morrison-Saunders A and Arts J (eds) *Assessing Impact: Handbook of EIA and SEA Follow-up* (London: Earthscan) 97–117.

Rosa, E and Short, J (2004) "The Importance of Context in Siting Controversies: The Case of High-level Nuclear Waste Disposal in the US". In Boholm Å and Löfsted R (eds) *Facility Siting: Risk, Power and Identity in Land Use Planning* (London: Earthscan) 1–20.

Sjölander-Lindqvist A (2004a) *Local Environment at Stake: The Hallandsås Railway Tunnel in a Social and Cultural Context* (Gothenburg: Centre for Public Sector Research, University of Gothenburg, and Human Ecology Division, Lund University).

Sjölander-Lindqvist A (2004b) "Visualizing Place and Belonging: Landscape Redefined in a Swedish Farming Community". In: Boholm Å and Löfsted R (eds) *Facility Siting: Risk, Power and Identity in Land Use Planning* (London: Earthscan) 107–126.

Sen, BP and Bhaskaran, TR (1962) "Anaerobic Digestion of Liquid Molasses Distillery Wastes", *Water Pollution Control Federation*, 34(10), 1015–1025.

Stoeglehner G, Levy J, Neugebauer G (2005) "Improving the Ecological Footprint of Nuclear Energy: A Risk-based Lifecycle Assessment Approach for Critical Infrastructure Systems", *International Journal of Critical Infrastructure*, 1(4), 394–403.

Stoffle B, Waters J, Abbott-Jamieson S, Kelley S, Grasso D, Freibaum J, Koestner S, O'Meara N, Davis S, Stekedee M, Agar J (2009) *Can an Island be a Fishing Community: An Examination of St. Croix and its Fisheries*. NOAA Technical Memorandum NMFS-SEFSC-593. (Miami, Florida: National Oceanic and Atmospheric Administration, National Marine Fisheries Service, Southeast Fisheries

Science Center). [Online] Available at: <http://www.stfavi.org/files/CananIsland beaFishingCommunity-Sept2009.pdf> (accessed 1 November 2011).

Stoffle R and Dobyns H (1983) *Nuvagantu: Nevada Indians Comment on the Intermountain Power Project.* Cultural Resource Series #7 (Reno, Nevada: Bureau of Land Management Nevada). [Online] Available at: <http://www.blm.gov/pgdata/etc/medialib/blm/nv/cultural/reports.Par.54831.File.dat/07_Nuvagantu,_NV_Indians,_Intermountain_Power_Project_1983.pdf> (accessed 1 November 2011).

Stoffle R, Dobyns H, Evans M (1983) *Nungwu Uakapi: Southern Paiute Indians Comment on the Intermountain Power Project.* Revised Intermountain-Adelanto Bipole I Proposal. Report submitted by Applied Urban Field School to Applied Conservation Technology. (Kenosha: University of Wisconsin-Parkside, Applied Urban Field School).

Stoffle R, Rogers G, Grayman F, Bullett-Bensen G, Van Vlack K, Medwied-Savage J (2008a) "Timescapes in Conflict: Cumulative Impacts on a Solar Calendar", *Impact Assessment and Project Appraisal,* 26(3), 209–218.

Stoffle R, Van Vlack K, Toupal R, O'Meara SN, Arnold R (2008b) *American Indians and the Old Spanish Trail. Prepared for the National Park Service, Trails Division* (Tucson, Arizona: Bureau of Applied Research in Anthropology, University of Arizona).

Stoffle R and Stoffle B (2007) "At the Sea's Edge: Elders and Children in the Littorals of Barbados and the Bahamas", *Human Ecology,* 35(5), 547–558.

Stoffle R and Minnis J (2008) "Resilience at Risk, Epistemological and Social Construction Barriers to Risk Communication", *Journal of Risk Research,* 11(1&2), 55–68.

Stoffle R, Toupal R, Zedeño N (2003) "Landscape, Nature, and Culture: A Diachronic Model of Human-Nature Adaptations". In: Selin H (ed.) *Nature Across Cultures: Views of Nature and the Environment in Non-Western Cultures* (Dordrecht: Kluwer Academic Publishers) 97–114.

Sutley N (2010) *Statement to the National Commission on the BP Deepwater Horizon Oil Spill and Offshore Drilling. Hearing August 25, 2010.* [Online] Available at: <http://www.whitehouse.gov/administration/eop/ceq/Press_Releases/August_25_2010> (accessed 1 November 2011).

Tollefson C and Wipond K (1998) "Cumulative Environmental Impacts and Aboriginal Rights", *Environmental Impact Assessment Review,* 18(4), 371–390.

5 Contested spatiality: geographical scale in sustainability assessment

Richard Howitt, Macquarie University

Introduction

Any development, plan, policy or action intended to pursue sustainability must address the question of scale. This is because securing human–nature relations that can be sustained into unpredictable futures requires consideration of institutional and administrative arrangements across ecological, social, economic and political structures that are themselves scaled. The idea of scale refers to the ways relations of governance, accountability and interdependence are played out across territory. The previous chapter focused on questions of temporal scale, and in this chapter the focus is on geographical scale and the practical challenges that arise from the observation that responses, solutions and approaches that make sense or can be effective at one scale are not necessarily effective or sensible at other scales (Fox, 1992). It is not just that places and landscapes are ecologically or sociologically different, but that different issues come into play as scale changes. Nor is it simply that in scaling up (for example from local to national to global) things get bigger. Indeed, the challenge of scale is not simply to deal with scale-as-size.

The complex and constantly changing interactions within and between elements of human and natural systems also demand consideration of scale-as-level and scale-as-relation (Howitt, 2003; Howitt, 1998). In other words, critical thinking about the geographical scales within which sustainability assessment frames the networks and connections that produce or constrain sustainability is essential to effective assessment. From the 1970s, issues of trans-boundary pollution and emergence of international jurisdictions in matters of environmental governance and human rights highlighted the need for attention to the international scale in many aspects of impact assessment. More recently, debates about both justice and sustainability, particularly arising from discussion of climate change adaptation (Adger *et al.*, 2005; Bulkeley, 2005; Smit and Wandel, 2006; Demeritt, 2001) have pushed decision makers to consider the ways that human actions affect social and ecological relationships across wider areas, and across the administrative and political boundaries created by human systems. Other pressures arising from the uneven spatial and social distribution of costs and benefits of major projects have added pressure to recognise scale as an important issue in environmental decision making and assessment (Macrory and Turner, 2002; Ogunseitan, 2003; Norman and Bakker, 2009; Hills and Roberts, 2001). Legacy effects for future generations from current decisions, and recognition that risks are associated with particular technologies and activities (long-term implications of non-renewable energy systems, long-term challenges of managing waste from nuclear fuel systems) (Fan, 2006; Sarewitz, 2004; Stoeglehner *et al.*, 2005; Stoffle and Arnold, 2003) have also pushed out the *temporal* scales at which the implications of decisions need to be considered by decision makers and the spatial boundaries that need to be encompassed in considering sustainability issues on greater temporal scales (see also Chapter 4).

Social movements and political institutions that address social and environmental injustice at a number of scales have also emphasised the importance of difference, interdependence and relatedness across social and cultural difference, and across human–non-human relations in ecological and social terms – what this chapter discusses as *social* and *ecological* scales – as well as more commonly discussed spatial boundaries and *geographical* scales. Such discourses push both regulators and practitioners of sustainability assessment towards accountable consideration of diverse others (other generations, other places, other species etc.) as central to the task of sustainability. Yet this is not a simple task that can be reduced to technical considerations or questions of method because defining appropriate reference points and preferred content of the idea of sustainability are hotly contested activities involving value judgements. The challenge of scale in sustainability assessment, therefore, is not solely methodological. It is also inescapably political and conceptual.

This chapter considers how students and practitioners of sustainability assessment might address the influence of geographical scale more thoughtfully, transparently and accountably. I argue that using scale as a framing concept offers powerful tools for exploring highly significant issues in sustainability assessment, as well as clarifying just what the basis (and limits) of any claim to sustainability might be. Rather than advocating any specific tool or set of tools, the chapter discusses how to respond to issues that move in and out of focus as the scale at which sustainability is considered changes. In the context of the idea of pluralism in sustainability assessment, the chapter suggests that sustainability assessment practitioners (and the decision makers they seek to inform) need to recognise that different social groups will privilege (and ignore) particular patterns of impacts as more or less relevant to the task of sustainability assessment. This situation creates a political context in which different scales come into play in assessing sustainability outcomes from the vantage points of different stakeholders. Ensuring that these scales are drawn into assessment and decision making is a key task for sustainability assessment teams. The chapter considers what actually constitutes data at various scales and in various circumstances, and discusses how relevant data might be identified, collected and interpreted in order to throw light on important issues affecting our understanding of the goal of societal and ecological sustainability.

Sustainability across geographical, social and ecological scales

In any impact assessment exercise, spatial and temporal boundaries need to be established to limit and prioritise which issues of impact, significance and connection are considered (Briassoulis, 1999; Gasparatos *et al.*, 2008). While it is true that natural and human systems are interconnected, and that to some extent everything is somehow connected to everything else, the task of impact assessment is to distinguish between significant and insignificant consequences. While the so-called 'butterfly effect', which suggests that the movement of a butterfly's wing (or in the original version from Ray Bradbury's 1952 science fiction story, the accidental destruction of a Jurassic butterfly by a human time-traveller) can have substantial consequences across vast spatial or temporal scales, is more conceptual than material, it nevertheless points to the need to consider cross-scale linkages. The importance of scale issues can also be glimpsed in the aphorism that someone cannot 'see the wood for the trees'. This delightfully ambiguous expression suggests that focusing on single trees stops an observer attending to the wider-scale issues of forest ecologies – or in a more commercial interpretation, one cannot focus on timber production if one is overly concerned with the management of individual trees. Such situations remind us that different – even contradictory – interpretation of even quite simple data is filtered through cultural and value systems to produce quite different versions of what makes sense; what is obvious, important or 'true'. It also reminds us that questions of scale are not simply a matter of 'size' or 'scope', but also implicate judgements about relationship and significance. In such circumstances singular, ostensibly authoritative representations of 'truth' are often inadequate responses to data that reflect multiple and simultaneous possibilities – each of which is 'true'.

Where to set the spatial and temporal boundary, how to describe the geographical focus and scope of any particular assessment in a globally connected system requires value judgements to be made at each step (Berger, 1981). Project developers, for whom impact assessment is often treated as a hurdle on the path to project approval, typically seek to disaggregate projects to allow consideration of smaller units of change in ways that disaggregate negative consequences. Colloquially this practice of fragmenting assessment is referred to as 'salami slicing', and Marsden (2011) highlights specific problems caused by projects which cross jurisdictional boundaries, with sub-projects being subject to assessment in each separate jurisdiction. The incremental impact of each component would have been minimised in such an assessment regime, and the individual components more likely to be approved (Maxwell *et al.*, 1997; McCutcheon, 1991; Salisbury, 1977; Vincent and Bowers, 1988).

Sustainability is, of course, a moving feast – the nature of social and ecological resilience (and vulnerability) means that responses to specific circumstances will change over time. Relationships that seem to promise the ongoing capacity to survive and thrive that marks sustainability can be affected by unanticipated perturbations such as an unexpected shift in population densities, climate or a specific event, and unpredicted consequences or cumulative burdens of change. For example, unusual fire or storm activities can substantially shift ecological relationships locally (Comaroff and Comaroff, 2001; Davis, 1995). Similarly, institutional arrangements that seem to hold promise of long-term progress can evolve into self-interested and maladaptive structures remarkably quickly (Gupta *et al.*, 2010; Jacobs and Mulvihill, 1995); and governance arrangements that are based on hard-fought negotiations and diplomacy can appear suddenly obsolete or inadequate in changing global circumstances (Holling, 2004; Bulkeley, 2005; Batterbury and Fernando, 2006; Liverman, 2004; Lockwood and Davidson, 2010).

To negotiate this moving feast, sustainability assessment tools must be developed that allow practitioners to work across and between scales in terms of ecological, social and organisational dynamics. It will not secure definitive answers that remove unavoidable uncertainties, but it should be able to identify key reference points for understanding, responding to and negotiating risks and inequities that emerge in the processes under assessment.

Sustainability in practice: cases studies of scale issues in sustainability assessment

Scoping and scale – assessment processes in Quebec (Canada)

In putting forward its proposals for development of the hydro-electric potential of Northern Quebec in the 1980s and 1990s, Hydro Quebec sought government agreement to have separate environmental assessments undertaken for the river regulation works (dams and pipelines), transport infrastructure (roads, airstrips and communication networks), power stations and power transmission lines. Locally affected communities in Northern Quebec and environmental advocates lobbied for a more comprehensive cumulative approach to the environmental assessment that acknowledged that the combined components constituted a single project, and the cumulative impacts needed to be considered in reaching judgements about the proposal. Environmental advocates pointed to the need for the spatial scope of the assessment to include the continental-scale effects on climate systems of changes the massive river regulation works would produce in runoff into James, Hudson and Ungava Bays and the likely changes in the timing of annual melting of sea ice in those areas, and the ecological consequences of such changes. Indigenous rights activists pointed to the need to ensure the spatial frame used by the assessment studies captured the complex human–nature interactions in so-called subsistence economies and the animals and ecosystems on which they depended. Thus one of the scale frames that was needed to ensure the assessment research supported the decision making required was the scale at which Indigenous land use, economic activity and governance was taking place. One of the difficulties in doing this is that while the Indigenous domain is often treated as 'local' in scale, it requires consideration of both ecological and social connections at much wider scales linked to larger spatial patterns of animal migration, seasonal environmental change and cultural interactions within Inuit and Cree populations, as well as wider political relationships around campaigns for Indigenous rights and interventions in political processes at provincial, national and international scales.

In developing assessment guidelines for this development, tribal, provincial and national authorities prioritised the need to make decisions that did not jeopardise economic relations that were already working consistently with principles of sustainable development (Evaluating Committee *et al.*, 1992). Ultimately the guidelines identified a small number of integrative concepts that required coherent consideration of human and natural systems, of multiple overlapping and conflicting value systems, and connections that created different sorts of geographical scope for different sorts of inquiries within the same assessment process. In this case, the assessment framework insisted that the scales at which the stakeholders themselves felt things mattered should determine the scope of the assessment process. So, rather than arbitrarily setting a pre-defined 'local', 'regional' or 'provincial' scope to decide what was included in or excluded from the study, the assessment guidelines required information be provided on the basis of the connections that were identified as important by the interested parties themselves.

The 'salami slicing' of projects into smaller elements for separate approval has both spatial and temporal consequences. The northern Quebec hydro-power schemes began with the James Bay (Le Grande) Project, which was planned as phase I of a program to regulate every river draining the northern Quebec plateau and involved construction of several large dams to impound reservoirs in forested areas. The project stimulated the negotiation of the *James Bay and Northern Quebec Agreement 1975* – the first of Canada's modern treaties with Indigenous peoples. One of the consequences of river regulation in this region was disruption of seasonal fisheries that provided an economic, nutritional and cultural foundation for many Indigenous communities. Not only would disruption of seasonal river flows affect the connections between the spring thaw of sea ice and the seasonal cycles of marine species, it would also have continental-scale consequences for climate processes. The impact assessment recognised such disruption was significant. To support the Indigenous economies, development of fisheries based on the reservoirs was proposed as a sustainable alternative (Berkes, 1988; Diamond, 1990; McCutcheon, 1991; Vincent and Bowers, 1988). An unanticipated impact of the inundation of forests was that timber rotting slowly in the cold waters of the reservoirs released methyl mercury into the riverine food chains. Both human and non-human predators faced significant threats from mercury poisoning from prolonged ingestion of contaminated foods. Once this problem was identified, the calamitous flaw in the original assessment and impact management strategies based upon it was clear. In this case, the cultural, economic and ecological effects of the contamination and its subsequent management were all outside the scope of the original decision making and the frameworks of accountability governing the decision makers designing and approving the development project were structured in ways that made it difficult to hold the decisions accountable at the geographical scales at which the problems emerged – the scales of Indigenous livelihood and governance.

This example points to the challenge of both predicting and negotiating life journeys – as individuals, communities and systems – 'whose path is unpredictable and unknown' (Holling, 2004, 1). As Stoffle *et al.* argue in Chapter 4 of this book, assessing sustainability effects across long time-horizons is almost impossibly difficult. Doing so in ways that relate to the various scales at which decisions affect social and ecological systems is equally difficult. It is essential that the sustainability assessment processes facilitate debate regarding what is sustainable in terms of social and cultural differences and across different ecological systems and also across different geographical scales.

Mining – bioaccumulation of radio nuclides Kakadu (Australia)

The ways in which scale boundaries affect the assessment of sustainability is well illustrated in the case of bioaccumulation of pollutants and transmission of negative effects along food chains which extend across the spatial and temporal boundaries commonly considered in terms of project-based assessments. Thinking along food chains requires different sorts of thinking about causation and in food chains that include long time-horizons and human consumption, questions of environmental justice and sustainable human–nature relations become significant.

At the Ranger Uranium Mine in north Australia's Kakadu National Park (Figure 5.1), significant scientific resources were marshalled to understand the environmental effects of the region's uranium mine. The Ranger mine is located inside a World Heritage Area listed for both its natural and cultural values. Understanding how high levels of local background radiation derived from the substantial uranium deposits affect natural and human systems, and how mining, processing and transporting the ore affects those systems, was

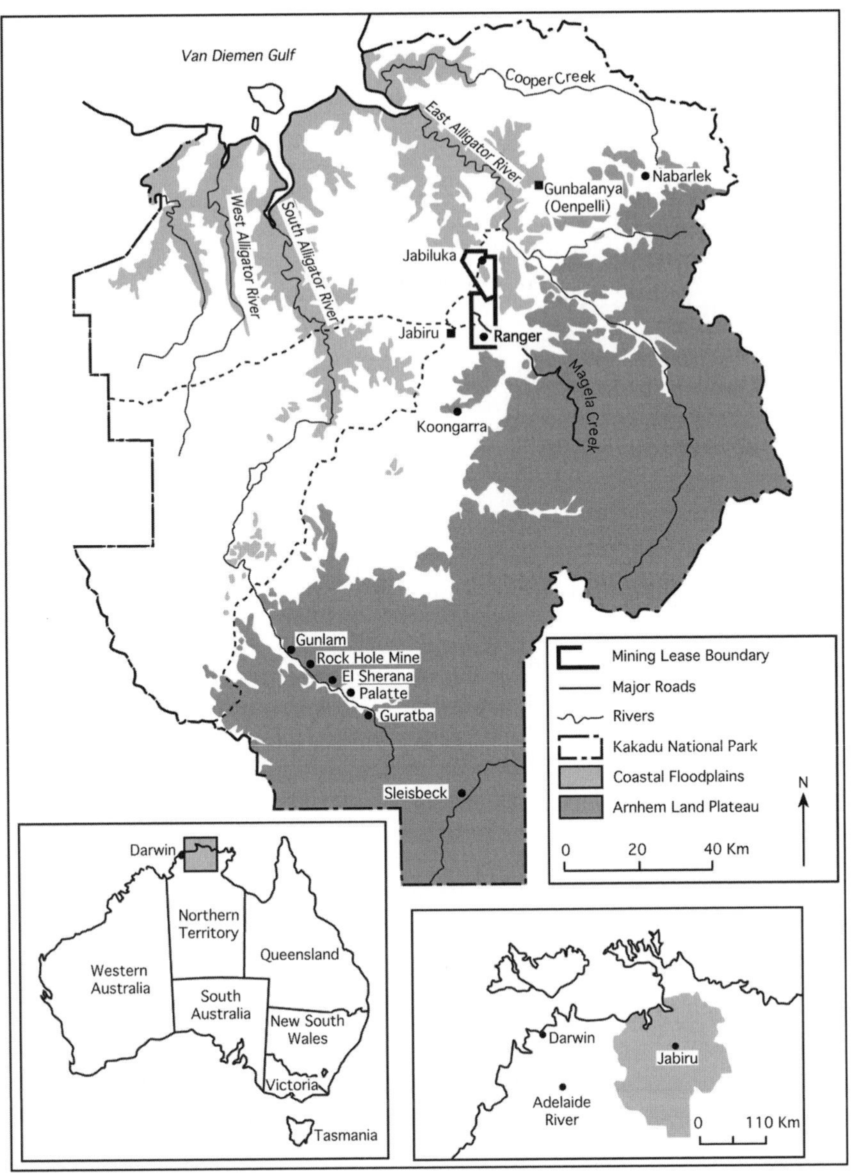

Figure 5.1

West Arnhem Region and the Ranger Uranium Mine, Northern Territory. Within the boundaries of Kakadu National Park, two mining leases (Ranger in the south and Jabiluka in the north) intrude into the conservation titles and community living areas, creating challenges for understanding and managing connections across boundaries and scales. (Drawn by Judy Davis, Department of Environment and Geography, Macquarie University) always central to environmental management in the region. Aboriginal people in the region faced different sorts of exposure risks to either the general population or the project workforce because they had different environmental relations, cultural practices and behaviours – including consumption of bush foods including fish, shellfish, turtle, birds and plants that create bioaccumulation risks (Tatz, 1984).

Valuable scientific research within the Kakadu region was intended to contribute to the management of the region's resources (Martin *et al.*, 1995; Martin *et al.*, 1997; Martin and Ryan, 2004). This work, however, was more concerned with expanding scientific knowledge than producing just and sustainable management outcomes (cf. Sarewitz, 2004) and paid little attention to communicating research results to the people who faced potential risks. The research provided detailed scientific discussion of the concentration of a number of radionuclides in a number of bush foods and identified high levels of radionuclide accumulation in some commonly used food species:

> Group 1 (bony bream and sleepy cod) had factors about five times higher than for Group 2 (eight other species including barramundi). Some smaller fish species (Group 3) are eaten whole and hence have relatively high concentration factors . . . Measured factors for fish in groups 1 and 3 were generally significantly higher than IAEA (International Atomic Energy Association) default values. Factors for turtle flesh were similar to those for the fish in group 1,but were about a factor of 10 higher for liver. Factors for magpie goose, filesnake, freshwater shrimp, goanna and crocodile flesh were also of the same order as for fish in groups 1 or 2.
>
> (Martin *et al.*, 1995: v)

Despite the scientific rigour of this research, it provided no discussion of the implications of its findings for Aboriginal health. The published papers from the research (Martin *et al.*, 1995; Martin *et al.*, 1997; Martin and Ryan, 2004; Ryan *et al.*, 2008) provide no explicit or easily accessible discussion, explanation or justification of their selection of species. There is no discussion of the nature of Aboriginal uses of these food resources, their cultural significance, or their place in the interface between the social and biophysical domains. No reference is made, for example, to Altman's fine study of outstation diets among Gunwinnggu speakers at Momega, about 140km east of the mine site, in similar environmental conditions (Altman, 1987). There was also no discussion of consultation with Aboriginal people about the methodology and implications of this research. The following comment (Martin *et al.*, 1995, 18) seems to exemplify the failure to prioritise Aboriginal concerns in the research: 'The large number of plant species used as Aboriginal food items made it impossible to include a comprehensive study as part of this project. Two root samples only . . . were analysed'.

While this research was conducted as part of a wide-ranging environmental monitoring process, it failed to integrate social and ecological concerns or to grasp the social, political and cultural relevance of its data to regional, provincial and national governance of the uranium production and waste management system. This was a remarkable oversight. It arises in large part from a poorly conceptualised scale frame, in which the scales of Indigenous livelihood and governance are devalued, and the international scientific community rather than the locally affected community becomes the preferred audience for reporting and responding to research findings. Dislocating the environmental data from its social and cultural context in this way reflects a localised myopia that simultaneously disconnects the environmental processes under examination from wider-scale ecological relationships (e.g. bird migrations and regional tidal flows) and longer time-horizons (e.g. intergenerational equity, child development concerns and longer-term institutional and governance arrangements). The reports in question here included detailed technical discussion of wider scientific research but linked none of this to discussion of the implications of this research for interested Aboriginal people. Again, the scale frame developed in the assessment defines a highly localised spatial scale as the scope for investigation and, while pursuing environmental data related to sustainability, twists the relationships and contexts around the data in ways that make social and cultural domains less visible and the broader sustainability implications unimportant.

By drawing spatial, temporal and conceptual boundaries in ways that took little or no account of what the appropriate geographical, social and ecological scales for the work might be, the research produced important documentation that was difficult for key stakeholders to make use of. In this sense, it exemplifies the challenge that faces researchers aiming to produce robust data for debate and decision in sustainability assessment processes. In this case, the persistent failure to consider the scales at which Indigenous peoples' connections to the social ecological systems of the Kakadu region were played out created a significant scale mismatch (Cumming *et al.*, 2006) between the decision-making requirements and the scientific reporting available to decision makers.

Murray-Darling Basin Authority (Australia)

Sustainable access to water as an economic, environmental, cultural and biological resource is a complex and compelling issue across the planet. It is widely identified as a driver of regional, national and international conflict (e.g. Feitelson, 2000; Hirsch and Wyatt, 2004).

Australia's largest river system, the Murray-Darling Basin (Figure 5.2), is important in economic, ecological, cultural and political terms and reflects many of the pressing issues affecting sustainable water management internationally. The basin covers one-seventh of the Australian land mass, including parts of Queensland, New South Wales, Victoria and South Australia, and the entire Australian Capital Territory. It has a human population of more than two million, and a total of three million people depend on water drawn

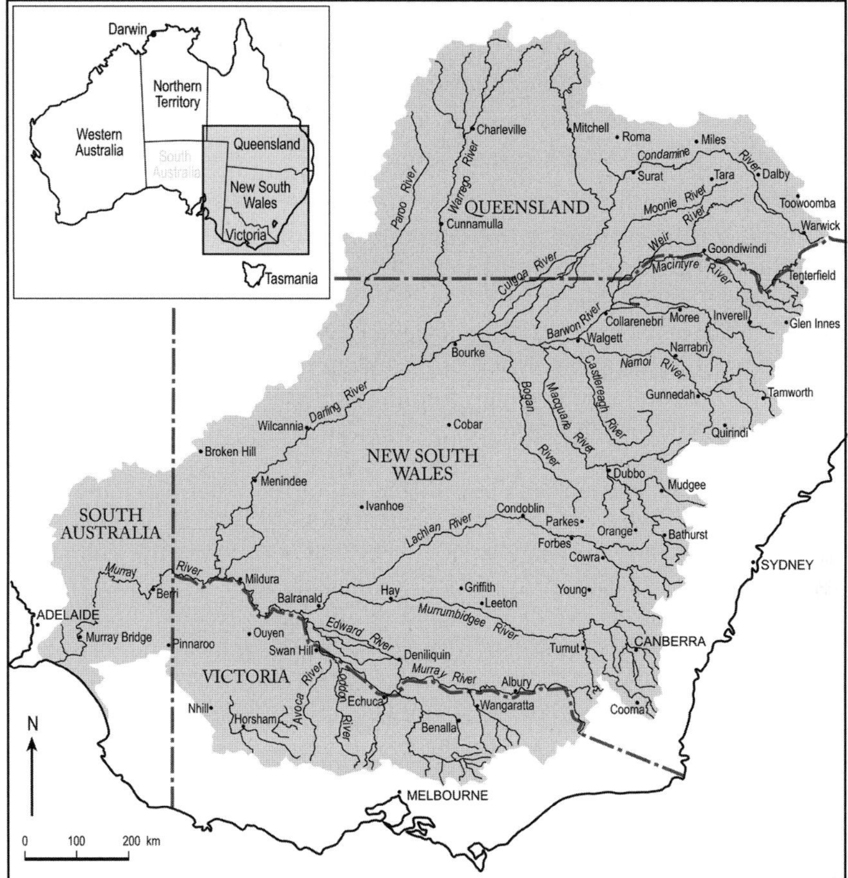

Figure 5.2

Murray-Darling Basin, Australia. Australia's largest river catchment includes four states and one territory, with widely diverse ecological, geomorphic and socio-cultural settings creating multiple scales for management within the catchment. Its links to state and national water politics and global agricultural markets creates another set of scale politics to be addressed in its management. (Drawn by Judy Davis, Department of Environment and Geography, Macquarie University)

from the basin. Agricultural activities within the basin produce over one-third of the nation's food supply including 53 per cent of Australian cereals grown for grain, 95 per cent of oranges, and 54 per cent of apples, and accounts for 65 per cent of the nation's irrigated land. The region is culturally diverse, with more than 30 Aboriginal nations with their traditional homelands in the region, and a high level of cultural diversity amongst settler communities as well. Ecological diversity is also high, with more than 30,000 wetland areas including internationally significant habitats for migratory birds and valuable heritage areas (Murray Darling Basin Authority, 2011).

While management of the basin's water resources was central to economic development from the time of European expansion into the inland, colonial and modern governments have struggled to balance competing demands and expectations. During periods of sustained drought from the 1980s to the 2000s, it became increasingly apparent that the balance of agricultural practices, urban and human consumption demands, and biodiversity and health requirements across the basin was unsustainable. In 1985 a joint ministerial council was established to discuss how governments could cooperate to achieve management goals at the geographical scale of the whole river basin. The basin was often at the centre of Australian debates about water reform, and the relationships between environment, economy and society. In 1992, the Murray-Darling Basin Agreement was enacted as a government–community partnership to 'promote and co-ordinate effective planning and management for the equitable, efficient and sustainable use of the water, land and other environmental resources of the Murray-Darling Basin' (Murray Darling Basin Commission, 2006). The administrative structure of this initiative comprised the Joint Ministerial Council as the decision-making body, the Murray Darling Basin Commission as the expert advisory and administrative body and a Community Advisory Committee, with representatives from a range of interest and expert groups.

Prolonged drought in 2002–2004 and heightened debate about questions of the quantity, quality and accessibility of water for economic, environmental and cultural uses, as well as deep concerns about the sustainability of water management systems in Australia in the context of emerging climate change, saw political pressure for change (Burton and Cocklin, 1996a; Burton and Cocklin, 1996b; Wentworth Group of Concerned Scientists, 2002). The National Water Initiative introduced a range of reforms, including new market mechanisms and creation of the Murray Darling Basin Authority (MDBA) to supersede the Commission in developing and implementing a Basin Plan for the entire catchment (McKay, 2005; Murray Darling Basin Authority, 2011). Following 18 months of technical deliberations and community consultations, the MDBA released its proposed Basin Plan in October 2010 (Murray Darling Basin Authority, 2009) to considerable public and academic debate (Connell and Grafton, 2011) and disagreements that saw the resignation of its chairperson. The Proposed Basin Plan is ambitious, tackling sustainable environmental, social and economic water management plans across political and administrative boundaries at the catchment scale. It is controversial because it challenges many powerful vested interests and historic assumptions about water resources in Australia. The challenges of working at the catchment scale across so many political, administrative and ecological differences are extreme, and the adversarial and self-interested political relationships across the basin mobilise scale in interesting and powerful ways. Indigenous peoples mobilise the scale of Indigenous territorial claims as 'country' to advocate management which prioritises connections that they referee demanding 'cultural flows'; irrigators simultaneously mobilise scales from the global market to local communities as scales that demand certain levels of flow to farms; environmentalists mobilise the scale of global bird migrations alongside the scale (temporal and spatial) of species extinctions and algal blooms to advocate specific environmental flows of water; and state and national politicians mobilise differently scaled representations of the public interest to support jurisdictional claims over decision-making pre-eminence. While planning in the Murray-Darling Basin has not been explicitly framed as a sustainability assessment exercise, concerns about sustainability of water resources, economic activities, communities and ecological assemblages have been central to the task of developing the Basin Plan. The complexly scaled politics involved inevitably represent a hurdle to securing sustainable outcomes and assessing the decisions and recommendations contained in the Basin Plan. Not least of these are the challenges of developing adaptive and innovative cross-scale institutions in a set of administrative structures that are characterised by fragmented and conflict-oriented administrative fiefdoms and interest groups.

Adger *et al.* suggest that even successful solutions in such settings see 'cross-scale linkages (that) evolve and are maintained by organizations and institutions . . . to further their own interests' (Adger *et al.*, 2005, 1). Transcending those interests to construct sustainable, innovative, adaptive and responsive institutional structures across communities of difference at various scales is a major challenge (Mulvihill and Keith, 1989; Berkes, 2002; Berkes, 2006; Cash *et al.*, 2007). Sustainability assessment practitioners operating in such institutional settings must include critical consideration of scale issues in their frameworks of practice.

Key ideas for sustainability assessment practice

One of the implications of scale in environmental governance is that 'neither purely local-level management nor purely higher level management works well by itself' (Berkes 2002, p.293), and that this means more than intervention at multiple scales that are treated as isolated from one another. The importance of working *across* scales is crucial. Wohling (2009) explores the extent to which Indigenous knowledges that are locally, temporally and culturally bounded can deal with environmental (and for that matter social, political and economic) changes that occur at greater-than-local landscape scales. In the context of rapid environmental change, developing governance and decision making that nurtures sustainability in complex ecological and human systems cannot rely on technological interventions to increase control and uncertainty (Pahl-Wostl *et al.*, 2009). Adaptive and resilient relationships and institutions across multiple scales are fundamental to sustainability. As the examples discussed above demonstrate, getting the scale wrong, or failing to facilitate adaptation and resilience across scales is destructive of sustainable outcomes. Sustainability assessment practitioners need to consider how the relationships and institutions they deal with respond and adapt to change at particular scales and across multiple scales.

Collecting data on resilience, adaptation and innovation is, of course, difficult and often subjective. Ensuring that sustainability procedures do not fall to accusations of naive subjectivism in such settings, the basis for making judgements and reaching conclusions, and the nature of the evidence relied upon, needs to be robust, coherent and defensible – even if it is subjective. Using social ecological systems analysis, Young suggests that environmental governance regimes 'have difficulty responding promptly and appropriately to socio-ecological changes that are non-linear and abrupt, even when the growing mismatch between prevailing institutions and the changing character of biophysical and socioeconomic systems becomes a matter of common knowledge' (Young, 2010, 384). The ways in which stresses producing change and challenge are formed at specific spatial and temporal scales; the ways in which they construct specific levels within social and ecological relationships; and the ways in which tipping points are produced and responded to are all issues that concern sustainability assessment.

To address this, sustainability assessment practitioners and regulators must establish the nature and drivers of change over a range of spatial scales, and across a range of ecological and social differences. Data that allows the nature of connectivities in terms, for example, of environmental services, vulnerabilities and resilience, accessibility and availability need to be transparently integrated into decision-making processes. But it also needs to be recognised that characteristics that can be identified and recorded at one scale (e.g. an institutional capacity to deliver a water quality intervention for a local community, and to secure water quantities for specific environmental and economic purposes within a locality) may be profoundly affected by processes constructed at other scales (e.g. drought or storm activity that reflects larger spatial and temporal systems than the locality and its administrative structures). Similarly, fragmenting sustainability assessments by selecting inappropriate spatial boundaries around ecological and social systems (as in the Canadian example discussed above), or by focusing at a single geographical scale (whether local, national or global) (as in the Australian examples discussed) is to risk missing critical data, or of misunderstanding the implications of data for key questions of sustainability.

This can be illustrated by returning to this chapter's initial linking of sustainability and justice. Environmental management regimes that seek to secure 'sustainable' management of water or any other natural resource by imposing injustices on specific localities or social or cultural groups inevitably face a risk of becoming ungovernable if groups bearing the burden of such injustice mobilise a scale politics that challenges the authority and solutions proposed. Indeed, the mobilisation of such scale politics is precisely what creates some domains as ungovernable. Thus, it is essential for practitioners, regulators and consumers of sustainability assessment not only to integrate thinking across multiple scales, but also to develop more flexible and sophisticated ideas about the ways in which geographical scale is drawn into debates about sustainability and the means to securing more sustainable futures.

References

Adger WN, Arnell NW, Tompkins EL (2005) 'Adapting to climate change: perspectives across scales', *Global Enviromental Change Part A*, 15, 75–76.

Adger WN, Brown K, Tompkins EL (2005) 'The political economy of cross-scale networks in resource co-management', *Ecology and Society*, 10. [Online] Available at: <http://www.ecologyandsociety.org/vol10/iss2/art9/> (accessed 24 November 2011).

Altman JC (1987) *Hunter-Gatherers Today: An Aboriginal Economy in North Australia* (Canberra: Australian Institute of Aboriginal Studies).

Batterbury SPJ and Fernando JL (2006) 'Rescaling governance and the impacts of political and environmental decentralization: an introduction', *World Development*, 34, 1851–1863.

Berger TR (1981) 'Public inquiries and environmental assessment'. In: Clark SD (ed.) *Environmental Assessment in Australia and Canada* (Melbourne: Ministry for Conservation) 377–400.

Berkes F (1988) 'The intrinsic difficulty of predicting impacts: lessons from the James Bay hydro project', *Environmental Impact Assessment Review*, 8, 201–220.

Berkes F (2002) 'Cross-scale institutional linkages: perspectives from the bottom up'. In: Ostrom E (ed.) *The Drama of the Commons* (Washington: National Academy Press) 293–321.

Berkes F (2006) 'From community-based resource management to complex systems: the scale issue and marine commons', *Ecology and Society*, 11, 45. [Online] Available at: <http://www.ecologyandsociety.org/vol11/iss1/art45/> (accessed 24 November 2011).

Briassoulis H (1999) 'Who plans whose sustainability? Alternative roles for planners', *Journal of Environmental Planning and Management*, 42, 889–902.

Bulkeley H (2005) 'Reconfiguring environmental governance: towards a politics of scales and networks', *Political Geography*, 24, 875–902.

Burton L and Cocklin C (1996a) 'Water resource management and environmental policy reform in New Zealand: regionalism, allocation and indigenous relations (Part I)', *Colorado Journal of International Environmental Law and Policy*, 7, 75–106.

Burton, L and Cocklin, C (1996b) 'Water resource management and environmental policy reform in New Zealand: regionalism, allocation and indigenous relations (Part II)', *Colorado Journal of International Environmental Law and Policy*, 7, 331–372.

Cash DW, Adger WN, Berkes F, Garden P, Lebel L, Olsson P, Pritchard L, Young O (2007) 'Scale and cross-scale dynamics: governance and information in a multilevel world', *Ecology and Society*, 11, 8. [Online] Available at: <http://www.ecologyandsociety.org/vol11/iss2/art8/> (Accessed 24 November 2011).

Comaroff J and Comaroff JL (2001) 'Naturing the nation: aliens, apocalypse and the postcolonial state', *Journal of Southern African Studies*, 27, 627–651.

Connell D and Grafton RQ (2011) *Basin Futures: Water Reform in the Murray-Darling Basin* (Canberra: ANU ePress).

Cumming GS, Cumming DHM, Redman CL (2006) 'Scale mismatches in social-ecological systems: causes, consequences, and solutions', *Ecology and Society*, 11, 14. [Online] Available at: <http://www.ecologyandsociety.org/vol11/iss1/art14/> (last accessed 24 November 2011).

Davis M (1995) 'Los Angeles after the storm: the dialectic of ordinary disaster', *Antipode*, 27, 221–241.

Demeritt D (2001) 'The construction of global warming and the politics of science', *Annals of the Association of American Geographers*, 91, 307–337.

Diamond B (1990) 'Villages of the dammed', *Arctic Circle*, 1, 24–34.

Evaluating Committee, Kativik Environmental Quality Commission, Federal Review Committee North of the 55th Parallel and Federal Environmental Assessment Review Panel (1992) *Guidelines: Environmental Impact Statement for the Proposed Great Whale River Hydroelectric Project* (Montréal: Great Whale Public Review Support Office).

Fan M-F (2006) 'Nuclear waste facilities on tribal land: the Yami's struggles for environmental justice', *Local Environment*, 11, 433–444.

Feitelson E (2000) 'The ebb and flow of Arab-Israeli water conflicts: are past confrontations likely to resurface?', *Water Policy*, 2, 343–363.

Fox J (1992) 'The problem of scale in community resource management', *Environmental Management*, 16, 289–297.

Gasparatos A, El-Haram M, Horner M (2008) 'A critical review of reductionist approaches for assessing the progress towards sustainability', *Environmental Impact Assessment Review*, 28, 286–311.

Gupta J, Termeer C, Klostermann J, Meijerink S, van den Brink M, Jong P, Nooteboom S, Bergsma E (2010) 'The adaptive capacity wheel: a method to assess the inherent characteristics of institutions to enable the adaptive capacity of society', *Environmental Science & Policy*, 13, 459–471.

Hills P and Roberts P (2001) 'Political integration, transboundary pollution and sustainability: challenges for environmental policy in the Pearl River Delta region', *Journal of Environmental Planning and Management*, 44, 455–473.

Hirsch P and Wyatt A (2004) 'Negotiating local livelihoods: scales of conflict in the Se San River Basin', *Asia Pacific Viewpoint*, 45, 51–68.

Holling CS (2004) 'From complex regions to complex worlds', *Ecology and Society*, 9, 11–20. [Online] Available at: <http://www.ecologyandsociety.org/vol9/iss1/art11/> (accessed 24 November 2011).

Howitt R (1998) 'Scale as relation: musical metaphors of geographical scale', *Area*, 30, 49–58.

Howitt R (2003) 'Scale'. In: Agnew J, Mitchell K and Toal G (eds) *A Companion to Political Geography* (Oxford: Blackwell) 138–157.

Jacobs P and Mulvihill P (1995) 'Ancient lands: new perspectives. Towards multi-cultural literacy in landscape management', *Landscape and Urban Planning*, 32, 7–17.

Liverman D (2004) 'Who governs, at what scale and at what price? Geography, environmental governance, and the commodification of nature', *Annals of the Association of American Geographers*, 94, 734–738.

Lockwood M and Davidson J (2010) 'Environmental governance and the hybrid regime of Australian natural resource management', *Geoforum*, 41, 388–398.

Macrory R and Turner S (2002) 'Cross-border environmental governance: the EC law dimensions', *Regional & Federal Studies*, 12, 59–87.

Marsden S (2011) 'Assessment of transboundary environmental effects in the Pearl River Delta Region: is there a role for strategic environmental assessment?', *Environmental Impact Assessment Review*, 31, 593–601.

Martin P and Ryan B (2004) 'Natural-series radionuclides in traditional Aboriginal foods in tropical northern Australia: a review', *The Scientific World Journal*, 4, 77–95.

Martin P, Hancock GJ, Johnston A, Murray AS (1995) *Bioaccumulation of Radionuclides in Traditional Aboriginal Foods from the Magela and Cooper Creek Systems* (Canberra: AGPS).

Martin P, Hancock GJ, Johnston A, Murray AS (1997) 'Natural-series radionuclides in traditional North Australian aboriginal foods', *Journal of Environmental Radioactivity*, 40, 37–58.

Maxwell J, Lee J, Briscoe F, Stewart A, Suzuki T (1997) 'Locked on course: Hydro-Quebec's commitment to mega-projects', *Environmental Impact Assessment Review*, 17, 19–38.

McCutcheon S (1991) *Electric Rivers: The Story of the James Bay Project (*Montréal: Black Rose).

McKay J (2005) 'Water institutional reforms in Australia', *Water Policy*, 7, 35–52.

Mulvihill PR and Keith RF (1989) 'Institutional requirements for adaptive EIA: the Kativik Environmental Quality Commission', *Environmental Impact Assessment Review*, 9, 399–412.

Murray Darling Basin Authority (2009) *The Basin Plan: A Concept Statement* (Canberra: Murray Darling Basin Authority).

Murray Darling Basin Authority (2011) *FAQs: About the Murray-Darling Basin* (Canberra: Murrray Darling Basin Authority, Australian Government).

Murray Darling Basin Commission (2006) *A Brief History of the Murray-Darling Basin Agreement [Archived website of MDBC]* (Canberra: Murray Darling Basin Commission).

Norman ES and Bakker K (2009) 'Transgressing scales: water governance across the Canada-U.S. borderland', *Annals of the Association of American Geographers*, 99, 99–117.

Ogunseitan OA (2003) 'Framing environmental change in Africa: cross-scale institutional constraints on progressing from rhetoric to action against vulnerability', *Global Environmental Change*, 13, 101–111.

Pahl-Wostl C, Sendzimir J, Jeffrey P (2009) 'Resources management in transition', *Ecology and Society*, 14, 46. [Online] Available at: <http://www.ecologyand society.org/vol14/iss1/art46/> (accessed 24 November 2011).

Ryan B, Bollhöfer A, Martin P (2008) 'Radionuclides and metals in freshwater mussels of the upper South Alligator River, Australia', *Journal of Environmental Radioactivity*, 99, 509–526.

Salisbury R (1977) 'A prism of perceptions: the James Bay hydro electricity project'. In: Wallman S (ed.) *Perceptions of Development* (Cambridge: Cambridge University Press) 172–190.

Sarewitz D (2004) 'How science makes environmental controversies worse', *Environmental Science and Policy*, 7, 385–403.

Smit B and Wandel J (2006) 'Adaptation, adaptive capacity and vulnerability', *Global Environmental Change*, 16, 282–292.

Stoeglehner G, Levy JK, Neugebauer GC (2005) 'Improving the ecological footprint of nuclear energy: a risk-based lifecycle assessment approach for critical infrastructure systems', *International Journal of Critical Infrastructures*, 1, 394–403.
Stoffle RW and Arnold R (2003) 'Confronting the angry rock: American Indians' situated risks from radioactivity', *Ethnos*, 68, 230–248.
Tatz C (1984) 'The social impact of mining: health, Aborigines and uranium hazards'. In: *Aborigines and Uranium: Consolidated Report on the Social Impact of Uranium Mining on the Aborigines of the Northern Territory* (Canberra: AIAS) 178–191.
Vincent S and Bowers G (1988) *Baie James et Nord Québècois: Dix Ans Aprés/James Bay and Northern Québec: Ten Years After* (Montréal: Recherches Amerindiennes au Québec).
Wentworth Group of Concerned Scientists (2002) *Blueprint for a Living Continent* (Geneva: Worldwide Fund for Nature).
Wohling M (2009) 'The problem of scale in indigenous knowledge: a perspective from Northern Australia', *Ecology and Society*, 14, 1. [Online] Available at: <http://www.ecologyandsociety.org/vol14/iss1/art1/> (accessed 24 November 2011).
Young OR (2010) 'Institutional dynamics: resilience, vulnerability and adaptation in environmental and resource regimes', *Global Environmental Change*, 20, 378–385.

6 Legal pluralism: notions of standing and legal process constraining assessment

Donna Craig, University of Western Sydney
Michael Jeffery, University of Western Sydney

Introduction

In a broad constitutional and governance sense, decisions, and decision-making processes, are ultimately controlled by courts. This chapter discusses the role courts play in the context of both common law and civil law jurisdictions in constraining sustainability assessment and explores the critical issue of legal standing which stipulates who has, and does not have, direct access to courts in relation to the sustainability assessment and the associated decision-making processes. Common law jurisdictions refer to those jurisdictions relying on that part of English law that is derived from customary use and judicial precedent together with statutory law, such as Canada, the United States, Australia and New Zealand whilst civil law (or civilian law) jurisdictions refer to those jurisdictions who have adopted a legal system inspired by Roman law, the primary feature of which is that laws are written into a collection (codified). Civil law is the dominant legal tradition today in most of Europe, all of Central and South America, parts of Asia and Africa, and even some discrete areas of the common-law world (e.g., Louisiana, Quebec and Puerto Rico) (Apple and Deyling, 1995). Parliamentary democracies depend on a separation of functions between the powers of the legislature (parliament making laws); the executive or administration (implementing laws) and the independent judiciary (enforcing the law) (Bates, 2010). It should be noted that this chapter focuses primarily on the role of the judiciary in countries with predominately democratic traditions. Therefore the ability for the courts and the legal system to influence sustainability assessment in countries such as China and North Korea will be more limited and the courts will not be seen to exercise decision-making powers independent of government.

This separation of powers is complex but fundamental to democratic governance. The reality is that separation of powers is rarely complete. The executive have limited powers similar to law making because of the devolution and discretion they have in implementing law and judges have always done more than merely 'interpret' law. The boundaries of separation remain important but it is more a 'balance' of powers. The concept of sustainable development, adopted throughout the world, since the United Nations Conference on Sustainable Development (UNCED, 1993) seriously challenges the assumption that the existing institutions and separation of powers is enough to deliver good governance. Much hope is placed on empowering civil society to direct this transformation, guided by *Agenda 21* (UNCED, 1993).

There are many theoretical and political assumptions implicit in this approach to social change and a heavy burden is placed on citizens and processes such as impact assessment. Similarly, law and judicial review cannot ensure sustainable decisions or good governance. Judicial review refers to the power of the courts to review administrative decision making with the historic origins of judicial review in common law jurisdictions being traced back to the ancient prerogative writs of mandamus, prohibition and certiorari. Mandamus compelled the performance of a public duty; prohibition prevented conduct outside jurisdiction and certiorari quashed past conduct for which there was no jurisdiction. In many jurisdictions the right to undertake judicial review has been extended by statute, for example, the Federal Court of Australia undertaking judicial review pursuant to the Administrative Decision (Downes, 2011). Courts in China, for example, do not have a general power of judicial review that enables them to strike down legislation. However, they do under the Administrative Procedure Law of the PRC have the authority to invalidate specific acts of the government such as the decision on administrative penalties and administrative permissions. That is not to say that sustainability assessment is of little importance in China, accounting for over one-sixth of the world's population and having recently surpassed the United States in carbon dioxide emissions, but rather an acknowledgement that the Chinese judiciary is not independent of government and cannot therefore play the same role as will be outlined below

A review of existing experiences with citizen access to justice and judicial review of impact assessment demonstrates the limits of what can be reasonably expected for future approaches supporting sustainability assessment. New international standards and approaches to governance provides some guidance about how future laws can be developed to further empower citizens and facilitate the transparency and review of significant decisions related to sustainable development. There is very little specific judicial review of explicit sustainability assessment, so this analysis needs to rely on analogy and experience in closely related areas.

The experience in jurisdictions influenced by British common law, such as the United Kingdom, Australia and New Zealand, is that there is considerable judicial restraint in overturning substantive merit and discretionary decisions of governments and delegated authorities. However, there is a robust tradition in environmental and administrative law related to procedural fairness and compliance. The judicial role is to ensure that decisions are made in the proper way and within legal powers. The reality is that a poor decision, on sustainability criteria, can be upheld as procedurally correct. An alternative approach is to embed sustainability in the constitutional framework and to provide broad standing for any citizen to bring an action to enforce them. An example of this approach is the national constitution of India and this will be briefly considered as an opportunity for more substantive judicial review.

Sustainable development and environmental impact assessment has become firmly embedded in the context of international environmental law and the focus of many of the global environmental challenges such as climate change, loss of biodiversity, protection of endangered species, food and water security lie in the international arena and are embodied in various principles, treaties and conventions, bilateral agreements and other initiatives negotiated between nation states. International treaties and conventions, when signed and ratified in accordance with the particular country's constitutional practice, must nevertheless be *implemented* by the contracting parties to be binding upon the individual citizen, and this requires the enactment of specific domestic legislation to be enforceable. It is for this reason that one must examine how sustainability assessment is approached at the national and state/provincial level of government and more particularly, the role that national laws and the judiciary play in interpreting sustainability assessment in the context of environmental decision making.

The detailed implementation of environmental obligations, enforcement, and review of decisions, remains largely within the jurisdiction (and environmental legislation) of each nation. This chapter will examine the opportunities for standing and judicial review of sustainability assessment in Australia, with a primary focus on implementation by the federal government (Australian Government) and the government of the State of New South Wales. The comments here will generally be applicable to other countries, although the specific methods adopted by particular jurisdictions with respect to implementation, enforcement and review of decisions will be tailored to fit that jurisdiction's planning and regulatory regime.

Context of sustainability assessment

Impact assessment, in many different forms and contexts (technology assessment, environmental impact assessment, social impact assessment, strategic and sustainability assessment) has a long and continuing history of evolution. Chapters 1 and 2 have argued the case for sustainability assessment, even though many experiences have undermined the rationalist view that better information and processes will improve quality of decisions based on them. There was a clear understanding in the original United States National Environmental Policy Act, 1969 (NEPA) that impact assessment was meant to inform, and transform, by integrating environmental values in decision making. There has largely been a failure to deliver beyond improved designs and mitigation measures. Public interest litigation (usually by non-governmental organisations (NGOs) and communities) has often been used to address this failure, usually challenging project-specific EIA (Jeffery, 2002).

The premise behind sustainability assessment is that potentially significant undertakings must be designed to deliver positive contributions to sustainability. This inextricably connects sustainability assessment to the related decisions and monitored outcomes. This is well understood as best practice but how can it be given legal definition and meaning? Another significant aspect of sustainability assessment is that it is inclusive of socio-economic factors as part of an integrated approach to the environment. This wider scope, and the concept of sustainable development, requires the equitable participation of affected communities and other stakeholders with different needs, perceptions and values. Thus sustainability assessment and related decisions should no longer be determined solely by 'science', experts or predetermined power relations and values, and the scope of assessment extends beyond particular projects. The modern approach is to identify sustainability assessment as part of good governance.

In terms of political theory, a variety of groups have a legitimate right to participate and assert their values. At the very least this is what is meant by pluralism. Many theorists believe that this does not go far enough to redress the existing imbalances of power (Craig, 1990). One response has been to develop better sustainability indicators and approaches to evaluating effectiveness (Bond and Morrison-Saunders in Chapter 3). Another dimension is to improve participation in sustainability assessment and related decision making, to improve the quality of the process, social learning and political effectiveness.

Sustainability assessment as part of improved governance: international initiatives

Environmental governance is evaluated by the effectiveness of strategies and initiatives implemented to achieve sustainable development and this has very significant normative and procedural aspects. These goals may increase capacity building, and increase access to environmental information, participation and justice. International soft law instruments (i.e. aspirational non-binding declarations, resolutions, statements of principle, codes of conduct etc.) such as the *Rio Declaration* (United Nations Conference on Environment and

Development, 1992), Agenda 21 (UNCED, 1993) and the World Conservation Union's (IUCN) Draft International Covenant on Environment and Development (IUCN Law Commission, 2010) set out the framework for achieving environmental goals such as these.

The United Nations Economic Commission for Europe (UNECE) Convention on Access to Information, Public Participation in Decision Making and Access to Justice in Environmental Matters, 1998 (Aarhus Convention) created a new kind of environmental agreement linking environmental rights and human rights. The Aarhus Convention (United Nations Economic Commission for Europe, 1998) originally applied primarily to the region of Europe but it has global significance for the promotion of environmental governance and now includes Canada and the United States. It also acknowledges that current generations owe an obligation to future generations and establishes that sustainable development can be achieved only through the involvement of all stakeholders in a democratic context.

The Aarhus Convention is built on three pillars: access to information; public participation in environmental decision making and access to justice in environmental matters. Environmental information is very broadly defined (Article 2(3)) and public authorities should make the information available to the public without requiring an 'interest' in the decision, without unreasonable charges (Articles 4(1) and (9)) and within a limited time (Articles 4(2), (3) (d) and (4)). There are exceptions but these should be subject to the Convention's provisions on access to review (Articles 4(7) and 9). Article 5 imposes innovative positive obligations on parties to 'possess and update' environmental information relevant to their functions and to establish systems to ensure an adequate flow of information to public authorities about activities that may have significant environmental impact (Articles 5(1) (a) and (b)). The public affected, or likely to be affected, have a right to participate in environmental decision making (Article 2(5)) and to have access to justice in order to challenge breaches of national law relating to the environment when their rights are impaired or they have 'sufficient interest' (Article 9).

In this context it is very relevant to sustainability assessment. It links government transparency, accountability and environmental protection and places public participation centre stage in sustainable development. Most importantly, it establishes that access to justice must be as broad as possible and be accompanied by rights to participate in decisions and reviews of them. This systematic approach challenges the older administrative law focus that prioritises procedural rights in judicial review. The national implementation of the Aarhus Convention throughout the European Community (Public Participation Directive 2003/35/EC, European Commission, 2006) has encountered some difficulties. However, it is now clear that environmental protection is a human right, with substantive content, that the overall policy objective is sustainable development and that public participation and civil enforcement processes are essential.

The procedural elements of Aarhus should not be implemented without having regard to the substantive obligations related to sustainable development. There is now an example of an international and national governance regime that inextricably ties these elements together. Sustainability assessment is one approach that can be used for improved environmental governance. The challenge for the courts is: who will be able to challenge sustainability assessment, on what basis (procedural and/or merit), what resources are available (Jeffery, 2002) and what remedies will be available to them?

International legal regimes provide normative standards and frameworks for implementation but offer limited opportunities for individuals and non-governmental organisations to challenge decisions of governments and corporations (with some exceptions in the field of regional and global human rights conventions). Some principles of sustainable development law are being included in many international treaties and declarations and are arguably becoming international customary law (i.e. law derived by customary use over time outside of treaty law and binding on all nations) (Birnie *et al.*, 2009). The principles of sustainable development are often included in the objectives of legislation but they are often not a mandatory standard used to evaluate, or review, decisions made under them. This is to be expected because of the difficulty in having mandatory requirements given the huge variety of contexts and factors taken into account. However, some significant steps could be taken in ensuring that these principles are necessary factors for consideration in all sustainability assessments and related decisions.

In relation to sustainability assessment, the precautionary principle is particularly significant. Where there are threats of serious or irreversible damage, lack of full scientific certainty should not be used as a reason for postponing measures to prevent environmental degradation (Peel, 2005). There is considerable ambiguity in some of these terms that has been discussed in Australian cases, but it has gradually become an accepted part of Australian environmental law. The most significant legal impact of the precautionary principle, when applied, is the effect on burden of proof. An objector to a proposal is usually required to provide the court with legitimate scientific evidence that raises the possibility of serious or irreversible environmental harm. Once this is established, then:

> the proponent would have to satisfy the burden of proof by evidence as to the likely consequences of the proposal, including scientific evidence (with its limitations) as to the proposed management regime and measures, and evidence to assist the court in the assessment of the risk weighted consequences of the proposal.
>
> (Bates 2010, p.199a).

It should be noted that the precautionary principle has application to a great many jurisdictions around the world including countries of the European Union, the United States and countries throughout Asia including China. It is one of the key international law principles around which there is growing opinion that its widespread adoption and use by many countries over a considerable period of time has placed it in the category of international customary law, and thus that it is binding, and this will undoubtedly be an issue for both national and international courts in coming years.

Standing and access to justice in Australia

The rules regarding *locus standi*, determining who has the right to bring an action before a court to enforce the law or highlight erroneous decisions, have traditionally been strict in Australia. The test of standing (or the right to appear and have party standing before the Court), under old common law, can be found in the case of *Boyce v Paddington Borough Council* which held that the plaintiff (the person bringing the action) must have suffered 'special damage peculiar to himself from the interference with the public right' (1903). The question of standing does not usually arise in the case of government (McGrath, 2008) and proponents of developments can establish private rights and interests that normally satisfy the test. The scope given to public interest litigants (those commencing the lawsuits in the public interest) to have standing to sue is a very important aspect of environmental governance, as it concerns the basic principles of participation and environmental justice. Special provisions have been enacted in environmental and planning legislation throughout Australia to improve citizen access to the courts.

The *Environment Protection and Biodiversity Conservation Act* 1999 (Cth) has now widened the scope for conservationists and conservation groups to seek judicial review, and obtain remedies, such as an injunction, to prevent breaches of the Act. Public interest litigants (people or bodies involved in conservation activity or detailed research) are given the power to seek an injunction on the grounds of false or misleading information being detected in an application for an approval or an Environmental Impact Statement (1999). Under the *Environmental Planning and Assessment Act 1979* (NSW), open standing provisions were incorporated to permit 'any person' to approach the court to seek to enforce any breach or apprehended breach of the law (1979). Since then, open standing provisions have been extended to most planning and environmental statutes in Australia.

This has been applauded as a positive move towards better environmental decision making, as environmentally committed citizens or groups have an avenue to voice their concerns in court. On the other hand, the limit under the Commonwealth legislation, that there must be false or misleading information, can be criticised as still too restrictive. Furthermore, loosened standing rules might mean there is increased potential for public participation, yet this participation can be described as almost meaningless, in the absence of the environmental advocates having sufficient resources to be able to present strong evidence and arguments. The lack of balance between proponents (in the public and private sector) and concerned citizen groups is not dealt with by changes to *locus standi* rules, but rather through the application of a court or tribunal's power to award costs, or through an integrated system providing for legal aid and/or intervenor funding (i.e. the funding of the opponents to the proposal or project before the court or tribunal) (Jeffery, 2002).

Regulatory authorities may enforce environmental laws by administrative orders (such as clean-up orders, environmental protection orders), by prosecuting offenders under criminal offence provisions and civil enforcement provisions. Civil and criminal enforcement can only be undertaken through court-based proceedings before a specialist environmental court or tribunal or the normal court structure (Bates, 2010). Failure to comply with an administrative order may result in civil or criminal enforcement. Criminal offences are defined under environmental legislation and liability extends to corporations, corporate officers, managers and employees (Protection of the Environment Operations Act, 1997). They are usually initiated by regulatory authorities and penalties such as fines and imprisonment can be imposed. This may be inappropriate when the purpose of an action, for breach of legislation, or common law duty, is to prevent environmental harm or to obtain compensation or remediation for environmental damage. In this situation civil enforcement can be brought by regulatory authorities and, in an increasing number of cases, by citizens and NGOs. Remedies such as injunctions, undertakings for damages, declarations and prerogative writs (such as mandamus, prohibition and certiorari) restrains a regulator from committing or continuing an illegal action (Bates, 2010). This area of law is heavily dependent on how defences are defined (to enable successful prosecutions) and the promotion of procedural compliance.

Challenging Australian environmental decisions

Merit appeals provide the only avenue in Australia to challenge the substantive quality, wisdom and fairness of a decision. Applicants for planning/ environmental approvals or licences under a permitting system to allow the project or proposal to go ahead usually have an opportunity for a merit appeal. This requires the appeal body to hear evidence and to uphold, amend or replace the original decision. The most significant opportunity for merit appeals is provided for objectors to designated development approvals, under the *Environmental Planning and Assessment Act 1979* (NSW), discussed below. This may involve a legal challenge to the substantive quality of an Environmental Impact Statement (EIS) required as part of the development application.

Current attempts of implementing sustainability assessment in jurisdictions such as NSW have been undertaken in the context of existing environmental impact assessment (EIA) law. Early attempts at sustainability assessments were often conducted as an addition to EIAs (Pope and Grace, 2006). The problem with this approach is that the shortcomings inherent in existing EIA processes have not been addressed and thus are likely to undermine the objective of sustainability assessments. In many respects attempts to incorporate and integrate sustainability assessment within the overall environmental assessment process have largely been unsuccessful. This is not surprising when one considers that EIA in Australia has from the outset adopted a project-specific methodology wherein environmental impacts are assessed and evaluated in the context of the specific project put forward for approval with little or no consideration given to the assessment of reasonable alternatives to the project itself.

EIA methodology, as conceived in the early 1970s, had the potential to include the assessment of policies, plans and programmes in addition to the assessment of a particular undertaking or project put forward by a proponent for approval. It offered an opportunity to break from the past and assess the environmental impacts of the proposed project in the light of not only all reasonable alternatives to it (including the nil option) but all reasonable alternative methods of carrying out the undertaking (i.e. implementing it (1980)). In the intervening years, however, most jurisdictions have moved away from this holistic concept of EIA as an essential component of the planning process and appear content to restrict its application to an assessment of the environmental impacts of the specific project under consideration (EPAA Part 4).

Thérivel has defined policies, plans and programmes in the following terms: policies are inspiration and guidance for action, plans are sets of co-ordinated and timed objectives for implementing the policy and programmes are a group of projects in a specific area (Thérivel, 2010).

Three principal benefits are often identified with integrating sustainability assessment into the policy, programme and plan phases of project development, namely, moving towards sustainability, strengthening the project EA and addressing cumulative and large-scale effects (Ouano, 2007). Sadler and Verheem noted as far back as 1996 that sustainability assessment (SA) is a process to incorporate environmental objectives and considerations into the upstream part of the decision-making process where traditionally economic, fiscal and trade policies guide the overall course of development (Sadler and Verheem, 1996). Sustainability assessment should therefore ideally be applied to policies, plans and programmes at the macro level before the project not after the fact. It is therefore important that sustainability assessment be introduced at the earliest opportunity in the planning process and not as an add-on to the EIA process itself, without running the risk of a seriously flawed EIA or the necessity of a costly reformulation of the project (Ouano, 2007).

Australian federal approaches to judicial review

The *Environment Protection and Biodiversity Conservation Act* 1999 (Cth) (EPBCA) is the main statute on which EIA is founded at the federal level. Here EIA applies mainly to identified actions requiring Commonwealth environmental approval. The main jurisdictional criterion is related to those actions that touch on matters of national significance including matters affecting Commonwealth actions and Commonwealth land (Lyster, 2009). It also introduces procedures for strategic assessment of policies, plans and programmes as well as actions. However, it is constrained by the application to matters of national significance, Commonwealth jurisdiction, and the consent of the environment minister (EPBCA Ch.4 Part 10, Division1) creating an ad hoc *voluntary process*. Subsequent EIA of projects may still be necessary but these are likely to be done under state and territory legislation with guidance from bilateral agreements. The *Intergovernmental Agreement on the Environment* came into effect in 1992 and co-operative federal approaches are adopted under the EPBC whereby the Commonwealth Government enters into assessment bilaterals to provide that Commonwealth concerns will be addressed in state or territory EIA. It is unlikely that strategic assessment under the EPBCA or other Commonwealth legislation will be amenable to civil actions such as judicial review by the public because it is voluntary and courts or tribunals will generally refuse to compel strategic assessment to be undertaken or evaluated unless it is a mandatory requirement of a statutory provision or regulation. Similarly, endorsement of a policy, plan or programme by a Minister is usually a political discretion that is not subject to judicial review.

Although there is still the 'special interest' test (first articulated in the *Boyce v Paddington* case referred to above) to overcome in the context of having access to the courts, including both the Federal Court of Australia and the High Court, in recent years the courts have significantly relaxed the application of the standing rules particularly in the context of litigation involving the protection of the environment (Wilson and Mckiterick, 2010). This is by no means the only barrier to using judicial review as a means of ensuring that sustainability assessment is incorporated in the planning or EIA process and, as is the case with all forms of litigation, the lack of funding and the discretionary power of the courts to award costs remain formidable obstacles.

New South Wales: state approaches to judicial review

EIA is required in New South Wales (NSW) for development that requires consent or approval or is to be undertaken by a public authority (Parts 3A, 4 and 5 of the EPAA; Lyster, 2009). The most appropriate level for SA would be in the preparation of environmental planning instruments (State Environmental Plan, Regional Environmental Plans and Local Environmental Plans) under Part 3 of the *Environmental Planning and Assessment Act (EPAA)*, 1979. Environmental studies may be required for new environmental planning instruments (EPIs) but most of the state is covered by existing EPIs. In this situation, the requirement for environmental studies to deal with strategic, cumulative, regional and local environmental impacts is largely a matter of government discretion. Any person (EPAA S.123) may bring an action (to the NSW Land and Environmental Court) to ensure that the requirements for making EPIs, under the EPAA, are complied with. This is one of the broadest possible 'standing' provisions, enabling any citizen to bring an action, but (in relation to EPIs) it only relates to procedural compliance.

Most EIA occurs, in the context of projects (Part 4 and 5 of the EPAA) requiring consent as designated development (Part 4) or an activity requiring a determination (Part 5) or a project 'called in' for ministerial determination (Part 3A). The EIA process under Part 5 is very similar to the Commonwealth EIA process except that the requirements for 'significance' and the contents of an EIA are included in legally binding regulations rather than guidelines (EPAR). The most significant avenue for substantive (merit) appeals of the decision (whereby the original decision can be replaced by the Land and Environment Court) is in relation to designated development under Part 4. Instead of leaving the issue of significant environmental impact to the minister, or other government authority, a list of developments is classified under the EPAA Regulations (EPAR Schedule 3). This is largely a list of extractive industrial and heavy polluting developments. All designated applications for development approval (consent) require an environmental impact statement that meets the specific requirement of the EPAA. An objector, who makes a submission on a designated development application within the specified time period, has a right to appeal to the Land and Environment Court on the merits of the development consent (and conditions).

This is a very significant advance in environmental law, as it connects the validity of an EIA to the decision-making process. Most importantly, the contents of an EIA are a legal requirement. This opened the opportunity for courts to review the substantive quality of EIA. This is a difficult task that has historically been approached with some reluctance by courts. The NSW Land and Environment Court is constituted by expert assessors as well as judges. Even in this situation, the courts are unwilling to impose too high a standard for the substantive content of an EIS.

Potential for legal review of sustainability assessment in Australia

Legal compliance and enforcement processes are relatively strong and well developed in the following circumstances:

- When legal duties are carefully defined and with enough precision to apply in specific contexts.
- When open standing enables citizen enforcement (predominantly through civil proceedings).

- When the remedy required involves procedural compliance with legislation or ensuring that the decision was made lawfully (within jurisdiction and power of decision maker).
- When adequate provision is made for cost orders in public interest citizen suits and for their funding.

Most of the legal review avenues apply to project level environmental assessment (including merit appeals against designated development in NSW). The implementation of a requirement for sustainability assessment, or invalidating a consent or approval because sustainability assessment was not undertaken, would require specific legislative provisions mandating sustainability assessment prior to a decision. Currently, this legal framework does not explicitly exist in Australia.

Learning from international and national approaches to improving environmental governance

The practice of sustainability assessment necessarily requires flexibility, environmental and cultural sensitivity and attention to the human rights and needs of disadvantaged people. Given this context, robust definition of legal duties and obligations will often undermine the purposes and best practice of sustainability assessment. However, procedural compliance is an essential element of lawful and transparent governance. The problem is that this is not enough to empower citizens, integrate sustainability assessment and values into decisions and to shift the burden of proof to require proponents of development to demonstrate that it will contribute to sustainability outcomes. Improving access to justice and judicial review cannot achieve this.

Sustainability assessment could be a part of a more holistic approach to improved environmental governance through:

- Requiring more systematic requirements for sustainability assessment of policies, strategies, plans, and programmes as well as projects.
- Implementing the requirements of the Aarhus Convention internationally and within nations.
- Legislating legally binding principles of sustainable development such as 'polluter pays' and the precautionary principle.
- Legislating a shift in the burden of proof relating to the application of sustainability principles to the party proposing an activity that may have sustainability risks.
- Legislating provision of funding for different forms of impact assessment by stakeholders and intervenor funding for procedural and merit appeals against related decisions.

Merit appeals will remain problematic. Ultimately, new ecological values and good decision making cannot be ensured through law. However, environmentalists and governments often neglect the use of law in the situations where it has proved robust (procedural compliance) and lack imagination in developing innovative approaches with the potential to empower and resource civil society in the inevitably political decision-making process.

Constitutional change and reform can expedite the task of legal and institutional change to facilitate sustainable development by entrenching environmental rights and extending existing human rights (such as the right to life). This has been incorporated into most constitutions of postcolonial and 'developing' nations such as the Philippines, South Africa and India. Their struggle against poverty, corruption and inequitable terms of trade in a globalised world often clouds the potential for the future implementation of these new constitutional rights. In India, M.C Mehta has used public interest litigation under the Indian Constitution to obtain landmark judgements and orders by the Supreme Court of India on environment and human rights over the last 25 years (Mehta, 2009). Hem Lata describes some of the most important and successful actions undertaken by Mehta to develop Indian jurisprudence for sustainability that promoted an activist court and required government action to redress problems and environmental damage (Lata, 2011).

In *M.C. Mehta v Union of India,* (also called the Oleum gas leak case), the court set aside the strict liability rule developed in the nineteenth century in *Rylands v Fletcher* in England in 1866 and ventured to develop a new rule of 'strict and absolute liability' to suit the particular conditions of the country. The court went on to say that judicial thinking should not be constricted by reference to the law as it prevails in England or any other foreign legal order and that courts should be prepared to receive light from whatever source it comes to build Indian jurisprudence. In *Indian Council for Enviro-Legal Action v Union of India,* the court accepted the polluter pays principle as law of the land and in the context of Indian environmental jurisprudence, it includes both the cost of compensating the victims of pollution and also restoring the environment to its undegraded condition.

In *Vellore Citizens Welfare Forum v Union of India,* the court adopted the precautionary principle into Indian environmental jurisprudence. The court broke the principle into three basic tenets: a) that the government and statutory authorities must anticipate, prevent and attack the causes of degradation; b) where there are threats of serious and irreversible damage, lack of scientific certainty should not be used as a reason for postponing measures to prevent environmental degradation and c) the onus of proof is on the actor or the developer/industrialist to show that their action is environmentally benign.

This case law was enabled by the persistence of a courageous Indian public-interest lawyer who understood that the constitution required that every person had a duty to uphold the law. The constitutional provisions, legal doctrines established and wide-ranging orders and remedies available in India are worthy of much further study to investigate and learn how far access to courts and judicial review can support the role of sustainability assessment in national regimes throughout the world.

Concluding comments

Although it is clear that courts have an important role to play in ensuring procedural and legal compliance their current role in the context of substantive and merit issues relies upon legislative or constitutional authority enabling the court to assume a decision-making role. The importance of procedural and legal compliance needs to be better understood in the context of separation of powers, particularly in a parliamentary democracy, including the extent to which it can ensure the accommodation of pluralism and the substantive aspects of sustainability assessment. Ultimately these aspects depend on a far more equitable, informed and resourced political process. Law reform can be useful in providing access to information, participation in decision making, broadening access to the courts and providing adequate funding for public interest litigation. To a large extent this is what is required to embrace national implementation of the international standards in the Aarhus Convention. Jurisdictions in many countries have made significant steps in this direction in recent years but are still deficient in key elements such as the provision of adequate funding and a lack of appropriate legal remedies.

References

1903, *Boyce v. Paddington Borough Council*, vol. 1 p.109.

1979, *Environmental Planning and Assessment Act (NSW)*, New South Wales, Australia.

1999, *Environmental Protection and Biodiversity Conservation Act (Cth)* in EPBCA, Commonwealth Australia.

Apple JG and Deyling RP (1995) *A Primer on the Civil-law System* (Washington DC: Federal Judicial Center).

Bates GM (2010) *Environmental Law in Australia*, 7th edn (London: LexisNexis Butterworths).

Birnie PW, Boyle AE and Redgwell C (2009) *International Law and the Environment*, 3rd edn (Oxford: Oxford University Press).

Craig, D (1990) 'Social impact assessment: politically oriented approaches and application', *Environmental Impact Assessment Review*, 10(1–2), 37–54.

Downes, Hon. Justice Gary (2011) *Judicial Review*. Seminar presentation for the College of Law, Government & Administrative Law, Sydney, 24 March 2011.

European Commission (2006) *On the Application of the Provisions of the Aarhus Convention on Access to Information, Public Participation in Decision Making and Access to Justice in Environmental Matters to Community Institutions and Bodies* in Regulation (EC) No 1367/2006, EPat Council.

IUCN Law Commission (2010) *Draft International Covenant on Environment and Development*, 4th edn (Gland, Switzerland: IUCN).

Jeffery MI (2002) 'Intervenor funding as the key to effective citizen participation in environmental decision making: putting people back in the picture', *Arizona Journal of International and Comparative Environmental Law*, 19(2), 35.

Lata H (2011) *PhD Thesis* (Sydney: University of Western Sydney).

Lyster R (2009) *Environmental and Planning Law in New South Wales*, 2nd edn (Annandale, NSW: Federation Press).

McGrath C (2008) 'Flying foxes, dams and whales: using federal environmental laws in the public interest', *Environmental and Planning Law Journal*, 25(5), 324–359.

Mehta M (2009) *In the Public Interest: Landmark Judgements and Orders of the Supreme Court of India on Environmental and Human Rights*, 3 vols (New Delhi: Prakriti Publications).

Ouano AR (2007) *Successful Pollution Control Through Cleaner Production, Myth or Reality?* PhD Thesis (Sydney: Macquarie University).

Peel J (2005) *The Precautionary Principle in Practice: Environmental Decision Making and Scientific Uncertainty* (Annandale: Federation Press).

Pope J and Grace W (2006) 'Sustainability assessment in context: issues of process, policy and governance', *Journal of Environmental Assessment Policy and Management*, 8, 373–398.

Sadler B and Verheem K (1996) *Strategic Environmental Asessment: Status, Challenges and Future Directions* (The Hague, The Netherlands: Ministry of Housing, Spatial Planning and the Environment).

Thérivel R (2010) *Strategic Environmental Assessment in Action*, 2nd edn (London: Earthscan).

UNCED (1993) *UNCED Report, A/CONF.151/26/Rev.1 (Vol.1) (1993)*, United Nations Conference on Environment and Development.

United Nations Conference on Environment and Development (1992) *Earth Summit: Rio Declaration & Forest Principles* (Rio de Janeiro: UNCED)

United Nations Economic Commission for Europe (UNECE) (1998) *Convention on Access to Information, Public Participation in Decision Making and Access to Justice in Environmental Matters* (Geneva: United Nations Economic Commission for Europe, Committee on Environmental Policy).

Wilson JD and Mckiterick M (2010) *Locus Standi In Australia – A Review Of The Principal Authorities And Where It Is All Going*. [Online] Available at: http://www.bawp.org.au/attachments/article/15/LOCUS%20STANDI%20IN%20AUSTRALIA%20-%20A%20REVIEW%20OF%20THE%20PRINCIPAL%20AUTHORITIES.pdf (accessed 24 November 2011).

7 Pluralism in practice

Jenny Pope, Integral Sustainability and North West University
Angus Morrison-Saunders, Murdoch University and North West University

Introduction

Drawing on examples from our experience we explore some of the dimensions of pluralism that typically manifest themselves in sustainability assessment practice. We build on the definition of pluralism provided in the Foreword to this book: 'the different interpretations which exist of a number of key issues relating to the outcomes of sustainability assessment' and other notions of sustainability assessment explored in previous chapters such as effectiveness, time horizons, spatial scales and legal processes.

The practice of sustainability assessment is pluralistic by definition, because sustainability itself is a contested concept (Davison, 2001). Sustainability is a broad concept that encompasses environmental, social and economic dimensions, with the result that different groups and individuals will have different views about the relative importance of these dimensions and the specific issues within them. Furthermore, because the practice of sustainability assessment is still in relatively early stages of evolution there are also alternative views of what sustainability assessment actually is and what it should be as a tool to promote sustainable decision making.

All of these issues arise in the conduct of almost any sustainability assessment process, usually manifested as differences in opinion of various stakeholders in the process. Much of this chapter therefore focuses on stakeholder and community engagement in sustainability assessment. It is important to note, however, that stakeholders internal to an organisation conducting a sustainability assessment are at least as important as the external stakeholders with whom the organisation engages. In both internal and external engagement processes, pluralism must be acknowledged, navigated and ultimately embraced. In this chapter we consider how to deal with pluralism in practice and also consider why pluralism is actually essential for effective sustainability assessment practice.

Context

There is a very broad spectrum, or pluralism, of sustainability assessment practice and processes around the world. The authors of this chapter live and work in Western Australia, and sustainability assessment practices in this particular jurisdiction are described in some detail in Chapter 10. However, since the examples we use to illustrate pluralism in practice come from our jurisdiction, it is worthwhile to briefly note here two forms of sustainability assessment that have been applied in Western Australia, both of which relate to projects rather than to planning or other more strategic forms of decision making:

- *External* sustainability assessment imposed by government on new projects proposed by private companies or government agencies for the purposes of determining whether or not a proposal should be approved and under what conditions.
- *Internal* sustainability assessment conducted by project proponents themselves as part of project planning, usually to select between options, which in turn are often alternative locations for infrastructure.

The first type of sustainability assessment is characterised as 'external' sustainability assessment because it is carried out in accordance with formal processes established by a body other than the proponent (in this case the government), while the second is 'internal' sustainability assessment because it is conducted by the proponent prior to any formal assessment process (Pope, 2006). The two types of sustainability assessment can also be distinguished by the decision question each is aiming to answer (Morrison-Saunders and Thérivel, 2006; Pope and Grace, 2006). The first is essentially a threshold question, 'Is this proposal sustainable enough?' (i.e. governments only wish to approve new development proposals that they consider to be acceptable or 'acceptably sustainable'), while the second is a choice question, 'Which is the most sustainable option?' (i.e. private companies and other proponents want to choose and proceed with the best alternative available to them).

The processes applied within each of these types of sustainability assessment in order to answer these questions are different, and pluralism manifests itself in correspondingly slightly different ways. In light of these types of sustainability assessment, we now explore some of the dimensions of pluralism in practice that we mentioned earlier, pluralism in conceptualisations of sustainability; pluralism in values and interests; and pluralism of process expectations.

Conceptualisations of sustainability

The contestability of the concept of sustainability has been widely acknowledged ever since the term sustainable development was popularised by the Brundtland Commission in 1987 (World Commission on Environment and Development, 1987). We will not go into this in detail in this chapter, except to note that different people's understanding of sustainability is generally related to their broader worldview.

For example, one person (perhaps the CEO of a mining company) may believe that sustainability means economic growth and development with appropriate environmental and social safeguards and mitigations being put in place, while another (perhaps a member of an environmental group) might believe that sustainability means protecting the Earth's natural systems and modifying society and/or the economy as necessary to achieve this imperative. Often this type of difference in viewpoint is manifested through weak and strong conceptions of sustainability (as outlined in Chapter 3). The key point we wish to make though is that it is unlikely that these two people will ever agree on a definition of sustainability, because their opposing views are grounded in very deep beliefs about the world and about life that typically don't change quickly.

What does this very fundamental example of pluralism mean for the practice of sustainability assessment? This can be particularly clearly illustrated by the first kind of sustainability assessment described above, in which the purpose of the sustainability assessment is to answer the threshold question, 'Is a proposal sustainable enough?' One example of such an assessment was the sustainability assessment conducted by the government of Western Australia in 2002–2003 to determine whether or not the Gorgon Gas Development should be allowed to be located on Barrow Island, a Class A Nature Reserve (akin to a national park, a Class A reserve has the highest conservation status in Western Australia requiring approval from both houses of Parliament in order to change its status such as enabling industrial development to occur within such a reserve). The Gorgon case study is discussed in more detail in Chapter 10, but the important point here is that this process essentially broke down completely as a result of pluralism in understandings of what sustainability means, especially in relation to management of the natural conservation values of Barrow Island.

The Gorgon sustainability assessment was facilitated by a working group composed of representatives of a number of government agencies, some of which were in favour of the development on economic grounds and some of which opposed it on environmental grounds. The process adopted was essentially to obtain and analyse as much environmental, social and economic information as possible, so that this group could determine whether or not the proposal was 'sustainable enough' and could make a recommendation to the government of Western Australia as to whether or not it should be approved. Unfortunately, however, no matter how much information was generated, members of the working group could not agree: the pro-development group members believed that the data demonstrated that the environmental risks associated with the development could be managed adequately to allow the project to go ahead so that the economic benefits could be reaped, while the pro-environment group members believed it demonstrated that the risks were too great.

In a very clear demonstration of pluralism in practice, in the end two separate reports written or commissioned by different state government agencies were sent to government, one recommending the development be approved, and the other recommending it be rejected. The government of Western Australia's own worldview apparently aligned most strongly with the pro-development group and the government determined that the development could be located on Barrow Island (Pope *et al.*, 2005). The proposal subsequently proceeded successfully to the next stage of assessment and decision making addressing the details of project construction and implementation. What was particularly interesting in this case with respect to pluralism is that different government agencies operating at the state level in Western Australia fell distinctly into the separate 'pro-development' and 'pro-environment' camps, whereas it might be expected that 'government' would represent a unified front.

Sustainability assessment is often described as an integrated process, whereby different considerations across a range of sustainability factors are (or should be) considered in a holistic way (Scrase and Sheate, 2002; Gibson *et al.*, 2005; Bond and Morrison-Saunders, 2011). Actually achieving integration is a constant challenge to the practitioner and the Gorgon example demonstrates how pluralism in a sustainability assessment process can undermine integration. However, we will also see later in the chapter that this was due in large part to the way in which this particular sustainability assessment was framed, and we will consider alternative ways of approaching sustainability assessment such that pluralism becomes an opportunity to *enhance* sustainability, rather than hinder it.

Values and interests

The Gorgon example demonstrates how different conceptualisations of sustainability are often underpinned by very different worldviews and values. This can be problematic when the purpose of the sustainability assessment is to determine if a proposal is sustainable enough because this question cannot be answered without a clear and agreed definition of sustainability (a point discussed in more detail by Bond and Morrison-Saunders, 2011). At the time of the Gorgon assessment, Western Australia did not have a definition of sustainability defined in policy. Even when there is a definition, it is usually so broad that there is still considerable room for different interpretations by people with different worldviews. This means that in a threshold decision there are usually winners and losers.

Many sustainability assessment processes, however, are of the second type introduced above, i.e. they involve a choice question and seek to determine which of a series of options is the most sustainable. Different stakeholders will still bring different values and worldviews to the table, but in choice decisions there are ways in which these can be incorporated into the assessment process so that compromises are possible. Two key dimensions of different stakeholders' value positions are the relative significance they place on different issues (for example across a broad sustainability agenda encompassing environmental, social and economic considerations) and what they consider to be acceptable and unacceptable with respect to these issues. We will consider how stakeholder values can be incorporated into sustainability assessments and decision-making processes in different ways in the following sections.

Multi-Criteria Analysis (MCA) for sustainability assessment

It is common when faced with a decision of choice to use decision-aiding techniques such as Multi-Criteria Analysis (MCA), which provide a transparent structure for the decision-making process. MCA techniques are particularly suitable for sustainability assessment processes based on choice because they allow a large number of considerations to be factored in to the decision, and explicitly enable the different value sets of diverse stakeholders to be incorporated (Bell *et al.*, 2003; Rauschmayer and Wittmer, 2006; Stirling, 2006; van den Hove, 2006). In general MCA processes work best when all of the options are broadly acceptable, i.e. none has a 'fatal flaw'.

The basic steps of an MCA process are:

- Development of options.
- Identification of sustainability criteria that form the basis of the decision (environmental, social and economic).
- Evaluation of each option against the criteria (scoring).
- Determination of the relative importance of each criterion to the decision at hand (weighting).
- Analysis (combining scores and weights to generate an overall performance value so that options can be ranked).

The values of different stakeholders are incorporated primarily in the weighting step of the process. There are many different ways in which weighting can be undertaken, but essentially it involves asking people to assign a numerical value to the importance a particular criterion has for a certain decision. For example, a stakeholder with strong environmental values may say that biodiversity is twice as important as visual impact, while someone living next door to a potential site for some new infrastructure might have exactly the opposite view. It is also important when conducting weighting to consider the impact that each criterion actually has on the decision. For example, biodiversity might be very important but if all the options have a similar impact on biodiversity, then this criterion may not be very significant to this decision, which is about choosing *between* options, notwithstanding that whichever option is chosen it would be imperative to manage biodiversity issues as best as possible.

MCA processes can be run in conjunction with a group of stakeholders or even a large community meeting. There are different ways of handling the weighting data obtained. Usually, every participant is asked to weight the criteria individually. These can then be averaged, or each person's weights can be run separately through the analysis step, or participants can self-nominate to represent a particular perspective (e.g. environmental, community, industry etc.) and the weights for all members of each perspective-group can then be averaged, run through the analysis and the results compared with results from other groups. In one Western Australian case study, stakeholders were invited to nominate the perspective with which they most identified (e.g. recreational groups, local residents, local businesses, environmental groups, operations and technical advisors) and the weighting was undertaken within each group and the results compared. As well as highlighting areas of difference, this process also found considerable common ground.

Working with stakeholders to determine their values and weighting of sustainability criteria can be very successful in our experience. However, there are some traps to be aware of. Often, a single stakeholder group will be polarised on an issue, with approximately half the people strongly favouring one option and half strongly favouring another, with each half identifying the other's preferred option as their least favoured option. In such cases, the MCA process may identify a 'best' (most acceptable) option as the one that no one hates, but no one likes either. A compromise has been struck between everyone's values, but everyone may be unhappy.

What's your bottom line?

Another way to approach a choice-based sustainability assessment is by a process of deliberation and negotiation around the 'bottom lines' or limits of acceptability perceived by different stakeholders. The purpose here is to identify whether any of the options are fatally flawed and to acknowledge that limits of acceptability are not usually hard and fast rules but need to be negotiated in the context of a particular decision situation.

For example, the major electricity utility in Western Australia was seeking to find a suitable site to build a new electrical substation to provide power to a new water treatment plant being constructed by the major water utility. The search area for the site has significant environmental and heritage values, is visited by significant numbers of tourists and day-trippers from Perth and is also an important water catchment. A large number of potential sites were investigated and many of these were ruled out for technical reasons, leaving a short-list of three.

As is usually the case in such situations, none of the three sites was ideal. Site 1 would have required the removal of a piece of industrial heritage, which although not a major tourist attraction was considered to be historically significant. Site 2 was well hidden from the road and therefore out of sight for tourists visiting the areas, but would have required the clearing of significant numbers of mature trees. Site 3 was in a cleared area adjacent to the major tourism attraction of the area and associated museum and potentially visible from a number of walking trails and view points in the area.

A stakeholder reference group was formed to discuss the site options and try to determine which one was best. The group comprised representatives of the local community, heritage agencies, environmental groups and others. The electricity utility also consulted with government regulators and agencies, including those responsible for environmental protection, water and heritage to seek their views. Although a wide range of environmental, social and economic issues were discussed, the three most important issues that emerged were visual impacts (since standard electricity substations are generally considered to be ugly), environmental impacts and heritage values. Each member of the stakeholder reference group and each government agency expressed their preferred site and why they believed some potential sites were unacceptable. Their preferences depended upon which of these issues they considered to be the most important, so some chose Site 1, some chose Site 2 and some chose Site 3. Discussions quickly reached a stalemate.

In this case it was decided not to undertake an MCA because it was not clear whether any of the sites were acceptable at all, i.e. the bottom lines or limits of acceptability were not clearly defined. The process therefore became about negotiating the limits of acceptability with the stakeholders, particularly in relation to the three key issues of environment, visual impact and heritage. The electricity utility invited representatives of agencies responsible for these key areas to a joint workshop. At the workshop, each representative had an opportunity to hear the point of view of every other representative. The group collectively agreed that the potential environmental impacts of developing Site 2 and the impacts on significant heritage of developing Site 1 were more unacceptable than the visual impacts associated with Site 3. It was agreed that the visual impacts could be mitigated by careful placement and screening of the substation and Site 3 was therefore selected as the preferred site by the group as a whole. Certain representatives would still have preferred that a different site had been selected but they could accept the group's recommendation.

This is an example of the use of deliberation in sustainability assessment to negotiate acceptability limits. We will return to consider deliberation and its role in sustainability assessment practice in more detail later in this chapter.

Process expectations

The examples in the previous section, whereby the sustainability assessment process aims to identify the most sustainable location for a development, are probably the most common forms of sustainability assessment currently practised in our jurisdiction of Western Australia, and are often undertaken by providers of public infrastructure such as water, electricity and roads and by local governments. The government of Western Australia also recently applied a similar approach to engage stakeholders and incorporate their values in the process of identifying a suitable site for a liquefied natural gas processing precinct in the north west of the state (see http://www.dsd.wa.gov.au/7909.aspx). There are also examples of large resource companies applying sustainability assessment to ensure that a broad range of sustainability considerations are factored into their site selection processes, as an important part of obtaining and retaining their social licence to operate (e.g. URS Australia, n.d.). These are discussed further in Chapter 10.

Community and stakeholder engagement is usually an essential part of these processes. While this is sometimes limited to a report being released for a period of public comment and review, it is also increasingly common for proponents to convene community fora or workshops in which participants are invited to participate in the sustainability assessment process in a variety of ways. There may be opportunities to comment on the options, suggest new options, identify pros and cons, or to participate in a structured MCA process, for example by contributing to the weighting process as discussed previously.

It is often the case, however, that some members of the community feel that the role offered to them is too limited and that the sustainability assessment itself is too constrained and may demand a larger, more influential role in planning and decision-making processes. These demands may often be in direct conflict with the views of some decision makers within the proponent organisation who prefer a more top-down, technical and expert-driven approach to decision making, with no community engagement at all. The sustainability assessment practitioner may therefore find himself/herself trying to work with a pluralism of views about the extent to which stakeholders and the broader community should be involved in decision making and sustainability assessment (see Fuller 1999, p.56 for a summary of typical stakeholder expectations in impact assessment).

One of the most common complaints of frustrated community members and other stakeholders is that the scope of the sustainability assessment is too narrow and that too many important, strategic decisions have already been made before the sustainability assessment commences. For example, in a sustainability assessment to determine the best route for a road, people may ask questions such as: Why does the road have to go in this area? Or even, why do we need this road at all? Has rail been considered? What about better public transport to encourage people not to drive?

It is also often pointed out that none of the options under consideration may really be sustainable (for example they may all have negative impacts on biodiversity), and that in fact the sustainability assessment is not about finding the most sustainable option but the least unsustainable one. There is thus pluralism in expectations of the sustainability assessment process (Scanlon and Davis, 2011).

In many cases, it is an unfortunate reality that key decisions that shape the sustainability assessment process will indeed have already been made, such as the need for the new road in the first place. These decisions may often be the result of planning processes undertaken many years ago, before sustainability was a consideration and before community engagement became common. For example, in the city of Perth, Western Australia, some of the roads currently being considered for construction first appeared on land-use plans formulated in the 1950s. One of these is currently the subject of a highly controversial assessment process restricted to identifying the best alignment of the road. Even if communities and stakeholders were engaged in these earlier, strategic planning decisions, the people involved were probably not the same people who are involved in the latest round of consultation and they don't necessarily have the same views.

The decision to build a new road may also simply reflect the mandate of the government agency responsible. For example a department of planning or a department of transport may be open to wide-ranging options for addressing mobility issues (e.g. railways, public transport, urban renewal and transformation) whereas a highways department will only consider road-related options, often involving construction of new roadways.

Plurality of expectations is a very real challenge for sustainability assessment practitioners. It is very important that expectations are managed as much as possible by clearly drawing the boundaries of the sustainability assessment, explaining the decisions already made and clearly defining the opportunities for involvement at the very commencement of the sustainability assessment. People then have a choice of contributing their views and values to the sustainability assessment at hand, or to oppose the project no matter which site or route is chosen, by challenging the decision making that has gone before through activism or lobbying. Ideally, though, sustainability assessment with meaningful community and stakeholder engagement would be part of strategic planning decisions as well as project-level decisions. We discuss this further in the next section.

Embracing pluralism in sustainability assessment

The previous discussion has focused predominantly on the challenges posed by various forms of pluralism in the practice of sustainability assessment. We have also alluded to ways in which pluralism can be managed, for example by clear framing and boundary setting, by more strategic applications of sustainability assessment and by deliberation, and each makes important contributions to learning and understanding. In this section, we take this further and explore these approaches in more detail, not just as ways to manage potentially difficult situations but ways to enhance the practice of sustainability assessment, and ultimately to contribute to better, more sustainable decisions.

Strategic approaches to sustainability assessment

In practice, the narrower the scope or frame of the sustainability assessment the more limited are the opportunities to deliver sustainable outcomes through the sustainability assessment process (Pope and Grace, 2006). Previously we posed two fundamental decision questions (threshold and choice) corresponding to external and internal sustainability assessment approaches respectively. However, for choice-type decisions, a spectrum of decision questions that vary in their level of 'strategicness' can be conceptualised (Morrison-Saunders and Thérivel, 2006; Hacking and Guthrie, 2008). The following example discusses this in relation to spatial planning and transport.

If the decision question is 'What is the most sustainable (least unsustainable) route for the new road?', the sustainability assessment will be limited to comparing the impacts of the different route options, most of which can be expected to be negative impacts (e.g. loss of biodiversity, noise impacts, congestion and air pollution, visual impacts etc.) and finding the 'least worst' option. It is likely that community forums will raise questions about the sustainability of cars as a form of transport, and spatial planning based on ever-expanding urban sprawl that requires more and more roads, but that they will be told that these issues are outside the scope of the sustainability assessment. It is also likely that pluralism in views will result in conflict that can only be resolved by someone winning and someone losing, or by compromises and trade-offs that leave everyone feeling cheated.

If, however, the question is framed as 'What is the most sustainable way to ensure an accessible city?' then the sustainability assessment process is quite different. The options identified could still include roads, but might also look at improved public transport, or high-density living where people can walk to amenities rather than driving. The opportunity for more innovative and sustainable outcomes is greatly enhanced. The process is likely to be far more positive, with participants all aligned with the common vision of a more accessible city. They will probably disagree on the best way to achieve this vision (pluralism will still be alive and well) but the process is likely to be more creative and focused on positive synergies rather than trade-offs between a range of unappealing options, with the overall outcome being far greater levels of community acceptance of the decisions.

Ideally, the framing or boundaries of the discourse would be established collaboratively whereby everything that is of concern or interest to the community is included and nothing is 'outside the scope' (Monnikhof and Edelenbos, 2001; Doelle and Sinclair, 2006). People are also more likely to participate in this broader sustainability assessment process if it is broadly and positively framed rather than deciding to oppose the proposal through other means. While this ideal may not always be attainable in practice, the key to success as a sustainability assessment practitioner is to push practice as high up the strategic spectrum as circumstances permit (i.e. avoid the trap of asking too narrow a decision question).

Deliberation in sustainability assessment

In the earlier example of the electricity substation we introduced the idea of deliberation in sustainability assessment. Deliberation is a highly contested term (Elster, 1998), much like sustainability itself, but in simple terms, deliberation means that a group of people, bringing a plurality of views to the table, 'carefully examine a problem and arrive at a well reasoned solution after a period of inclusive, respectful consideration of diverse points of view' (Gastil and Black, 2008, p.2).

In our example, the deliberative process was quite modest, with just a few key stakeholders in the room, and its purpose was to enable each stakeholder to understand the positions and views of the others, to determine which bottom lines or acceptability limits could be negotiated. While this proved a useful exercise that achieved its purpose, the potential of deliberative approaches to sustainability assessment extends much further into an exploration not just of different views and values, but the beliefs (or worldviews) that underpin them (Gundersen, 1995). This process is sometimes also called 'dialogue' following the ancient Greek tradition (Roberts, 2002).

The Western view of decision making leans towards open, transparent and participative democratic processes, as espoused in international best practice principles for impact assessment such as IAIA (1999). Deliberation or dialogue is at the heart of the notion of deliberative democracy, which has a political dimension and also emphasises the empowerment of the general public in decision making. The tenets of deliberative democracy have been described (Hartz-Karp and Pope, 2011) as:

- Representativeness, meaning that the people deliberating and making decisions are ordinary people representative of a society's demographic (who are likely to bring a pluralism of views).
- Deliberativeness, meaning that they engage in deliberations as we have defined above, that respect and explore plurality.
- Influence, meaning that the decisions reached by this group have real influence over policy or decision making.

Deliberative democratic approaches to sustainability assessment are potentially better equipped to embrace pluralism than other more technical approaches such as MCA discussed earlier.

Even without all the features of deliberative democratic approaches such as representativeness or influence, deliberation or dialogue can be of enormous value when applied as part of a sustainability assessment process. It can facilitate the resolution of issues and the emergence of new ideas or solutions to challenging decision-making situations. This was demonstrated in another sustainability assessment undertaken in Western Australia, the assessment of the South West Yarragadee (SWY) Water Supply Development conducted in 2004–2006.

Very simply, the proposal here was to extract 45 GL/day of groundwater from the Yarragadee aquifer in the south west of the state and pump it 300km to the capital city of Perth to supplement the integrated water supply (Strategen, 2006). The proposal was opposed by small communities in the south west which were not connected to the integrated system and had to rely on small town water supplies that were often unreliable in summer. Their concern was the loss of potential future uses for the water for private initiatives (e.g. an orchard or a dairy) in their own region.

The sustainability assessment process for the SWY proposal was conducted before the proposal was finalised, which meant that there was an opportunity to refine the proposal through the sustainability assessment to make it more sustainable. The process was coordinated by a working group comprising the proponent and its consultants who were responsible for environmental, social and economic studies of the proposal. Pluralism was evident amongst the group from the beginning, and the views of the social and economic consultants were particularly opposing. The economic analysis showed that the best economic use of water was within an integrated system, which in terms of the SWY proposal meant the economic analysis favoured the proposal to send the water to Perth (the capital of Western Australia 300km to the north). On the other hand, the qualitative social impact assessment highlighted the sense of 'futures foregone' experienced by the local communities in the south west, which was of concern to the decision makers and perceived to indicate a lack of social sustainability of the proposal.

The members of the working group all brought very different backgrounds and perspectives on the issue but were willing to talk and try to understand each other's views and the underlying reasons for these views. Through this process of deliberation it became clear that there was a solution that would meet both the social and the economic objectives, and that was to extend the integrated water supply system to the south west towns at the same time that the SWY aquifer was developed. Therefore, the economic value of the water would be maximised and the local communities would have a reliable supply of water and not feel that this was being unfairly taken away from them. In this way, the sustainability assessment for the SWY proposal integrated competing considerations in a way that the Gorgon assessment did not, and the result was a more sustainable proposal, which all members of the working group supported.

We believe that deliberation can therefore not only resolve conflicts that can arise from pluralism in sustainability assessment, but can be generative, facilitating the creation of more sustainable proposals than originally conceived. But even more fundamentally, we believe that deliberation about sustainability in sustainability assessment processes can facilitate deep learning that extends beyond just the decision at hand. We explore this in the following section.

Sustainability assessment as a learning process

There are many different kinds of learning that can occur through sustainability assessment processes. Environmental, social and economic studies are typically undertaken and this information is provided to the decision makers.

Since the information is often new, the decision makers learn something by reading it. This practical type of learning has been called instrumental learning (Sinclair *et al.*, 2008). The type of learning that occurred within the project team in the SWY case, whereby the concept of the proposal was redefined, went beyond instrumental learning and could be considered a form of conceptual learning (Glasbergen, 1996).

However, learning can also be even more profound than this. In their analysis of policy learning in EIA, Sinclair and Diduck (2001) use the idea of 'adult transformative learning' that occurs through deliberation or dialogue as we have described them, when people representing a plurality of views come together to address a problem or a decision. This can manifest itself not just in a new way of thinking about the decision at hand, as occurred in the sustainability assessment process for the SWY proposal, but a new way of thinking about sustainability and about the world.

We believe that sustainability assessment provides the perfect situation to promote this kind of *transformative learning*, precisely because of the pluralism that is so much a part of the character of sustainability assessment. In fact sustainability assessment can be described as a process of decision making through pluralism. Sustainability as the concept at the heart of sustainability assessment brings with it ambiguity and inherent tensions, and each decision-making situation brings these into sharp relief within a particular context. Sustainability assessment requires multi-disciplinary teams, whose members bring plurality of expertise, backgrounds and beliefs to the process. In a respectful, non-confrontational environment, it is the plurality itself that provides the tensions that allow beliefs to be challenged and creative solutions to emerge.

The potential of sustainability assessment as a forum for learning through plurality is perhaps summed up best by a government participant in the SWY process, who said:

> This process has started me thinking more deeply about some of those bigger issues . . . it's just made me think differently about it and more deeply. If it can achieve that for me, just working on one little project in our little corner of the world, if we can encourage more of it on other big initiatives and strategic projects and get more and more people involved in looking at things that way, it is going to have an impact.
>
> (Pope 2007, p.269)

Conclusion

Pluralism lies at the very heart of sustainability assessment practice. As practitioners, we must not just acknowledge pluralism but embrace it as an essential aspect of our practice. The key learnings arising from our experiences with sustainability assessment that future practitioners might find useful are:

- Pluralism in the practice of sustainability assessment arises in many forms, including those we have discussed in this chapter, i.e. conceptualisations of sustainability, values and interests, and process expectations.
- Understanding of pluralism should inform the design of sustainability assessment processes to ensure that expectations are managed and that different views and values are reflected in the process and the decision.
- Decision questions that are as open and strategic as possible will maximise constructive stakeholder participation and the acceptance of decision outcomes by a range of stakeholders, as well as delivering more sustainable outcomes.
- Deliberative sustainability assessment processes, based upon respectful dialogue and a willingness to challenge and explore beliefs and assumptions, can facilitate not just better decisions but transformative learning for sustainability.

References

Bell ML, Hobbs BF, Ellis H (2003) 'The use of multi-criteria decision-making methods in the integrated assessment of climate change: implications for IA practitioners', *Socio-Economic Planning Sciences*, 37, 289–316.

Bond AJ and Morrison-Saunders A (2011) 'Re-evaluating sustainability assessment: aligning the vision and the practice', *Environmental Impact Assessment Review*, 31(1), 1–7

Davison A (2001) *Technology and the Contested Meanings of Sustainability* (Albany, USA: State University of New York Press).

Doelle M and Sinclair AJ (2006) 'Time for a new approach to public participation in EA: promoting cooperation and consensus for sustainability', *Environmental Impact Assessment Review*, 26(2), 185–205.

Elster J (1998) 'Introduction'. In: Elster J *Deliberative Democracy*. (Cambridge, UK and New York: Cambridge University Press).

Fuller K (1999) 'Quality and quality control in environmental impact assessment'. In: Petts J (ed.) *Handbook of Environmental Impact Assessment. Volume 2 – Environmental Impact Assessment in Practice: Impact and Limitations* (Oxford: Blackwell Science), 55–82.

Gastil J and Black LW (2008) 'Public deliberation as the organizing principle of political communication research', *Journal of Public Deliberation*, 4(1), Article 3.

Gibson R, Hassan S, Holtz S, Tansey J and Whitelaw G (2005), *Sustainability Assessment Criteria, Processes and Applications* (London: Earthscan Publications).

Glasbergen P (1996) 'Learning to manage the environment'. In: Lafferty WM and Meadowcroft J (eds) *Democracy and the Environment: Problems and Prospects*. (Cheltenham, UK: Edward Elgar), 175–193.

Gundersen AG (1995) *The Environmental Promise of Democratic Deliberation* (London: University of Wisconsin Press).

Hacking T and Guthrie P (2008) 'A framework for clarifying the meaning of triple bottom-line, integrated, and sustainability assessment', *Environmental Impact Assessment Review*, 28(1), 73–89.

Hartz-Karp J and Pope J (2011) 'Enhancing the effectiveness of SIA through deliberative democracy'. In: Vanclay F and Esteves, AM (eds) *New Directions in Social Impact Assessment: Conceptual and Methodological Advances* (Cheltenham, UK: Edward Elgar Publishing Limited).

Monnikhof RAH and Edelenbos J (2001) 'Into the fog? Stakeholder input in participatory impact assessment', *Impact Assessment and Project Appraisal*, 19(1), 29–39.

Morrison-Saunders A and Thérivel R (2006) 'Sustainability integration and assessment', *Journal of Environmental Assessment Policy and Management*, 8(3), 281–298.

Pope J (2006) 'Editorial: what's so special about sustainability assessment?', *Journal of Environmental Assessment Policy and Management*, 8(3), v–ix.

Pope J (2007) *Facing the Gorgon: SA and Policy Learning in Western Australia*, PhD thesis, Murdoch University. [Online] Available at: <http://researchrepository.murdoch.edu.au/264/> (accessed 7 July 2011).

Pope J and Grace W (2006) 'Sustainability assessment in context: issues of process, policy and governance', *Journal of Environmental Assessment Policy and Management*, 8(3), 373–398.

Pope J, Morrison-Saunders A, Annandale D (2005) 'Applying sustainability assessment models', *Impact Assessment and Project Appraisal*, 23(4), 293–302.

Rauschmayer F and Wittmer H (2006) 'Evaluating deliberative and analytical methods for the resolution of environmental conflicts', *Land Use Policy*, 23(1), 108–122.

Roberts NC (2002) 'Calls for dialogue'. In: Roberts NC (ed.) *The Transformative Power of Dialogue* (Greenwich, CT: JAI Press), 3–24.

Scanlon J and Davis A (2011) 'The role of sustainability advisers in developing sustainability outcomes for an infrastructure project: lessons from the Australian urban rail sector', *Impact Assessment and Project Appraisal*, 29(2), 121–132.

Scrase I and Sheate W (2003) 'Integration and integrated approaches to assessment: what do they mean for the environment?', *Journal of Environmental Policy and Planning*, 4(1), 275–294.

Sinclair AJ and Diduck AP (2001) 'Public involvement in EA in Canada: a transformative learning perspective', *Environmental Impact Assessment Review*, 21(2), 113–136.

Sinclair AJ, Diduck A, Fitzpatrick P (2008) 'Conceptualizing learning for sustainability through environmental assessment: critical reflections on 15 years of research', *Environmental Impact Assessment Review*, 28(7), 415–428.

Stirling A (2006) 'Analysis, participation and power: justification and closure in participatory multi-criteria analysis', *Land Use Policy*, 23(1), 95–107.

Strategen (2006) *South West Yarragadee Water Supply Development: Sustainability Evaluation/Environmental Review and Management Programme (ERMP)*. Report prepared for Water Corporation, Perth (Subiaco: Strategen).

URS Australia Pty Ltd (n.d.) *Pilbara LNG Project Site Selection Study* (Perth: BHP Billiton).

van den Hove S (2006) 'Between consensus and compromise: acknowledging the negotiation dimension in participatory approaches', *Land Use Policy*, 23(1), 10–17.

World Commission on Environment and Development (1987) *Our Common Future* (Oxford: Oxford University Press).

Part 3

Sustainability assessment: practice

8 Framework for comparing and evaluating sustainability assessment practice

Alan Bond, University of East Anglia, Angus Morrison-Saunders, Murdoch University and North West University, Richard Howitt, Macquarie University

Introduction

The aim of this chapter is to develop a framework to compare and evaluate the effectiveness of sustainability assessment practice in different jurisdictions. To do this, it is important to clarify what is meant by effectiveness. Chapter 3 set out a typology of effectiveness criteria derived from the academic literature and identified that effective sustainability assessment involves procedural, substantive, transactive and normative elements. The key message from Chapter 3 is that effectiveness is difficult to measure in absolute terms because of the diverse and even divergent reference points against which effectiveness might be judged. Consequently, in comparing and evaluating sustainability assessment in different places, the way that the ecological, social, political and cultural pluralism that provides the context in which the work of sustainability assessment is done must be recognised and accommodated as a central point of any comparative discussion.

The framework outlined in this chapter considers how this emerging field of practice integrates learning and knowledge into continuous improvement (Boothroyd *et al.*, 1995; Jha-Thakur *et al.*, 2009). Figure 8.1 depicts the four categories of effectiveness introduced in Chapter 3, and incorporates the critical influences of pluralism, and knowledge and learning, into a typology that provides a coherent framework for comparative evaluation of sustainability assessment across different jurisdictions, times and approaches in terms of methods and data availability. We recognise that sustainability assessment is a relatively new practice, and that like any field of professional practice, it needs constant review, development and improvement within the community of practice. Therefore it is inappropriate to attempt to create here a single hard-and-fast set of criteria to compare and evaluate effectiveness. Rather, this section explores how both the formally prescribed operation of sustainability assessment (equivalent to external sustainability assessment as defined in Chapter 7), and the less formal cultures of professional practice (equivalent to internal sustainability assessment as defined in Chapter 7) are evolving in different places, and what that experience brings to the task of improving future sustainability assessments.

Based on this typology, it is possible to propose a framework for comparing sustainability assessment processes in different settings to guide expert judgement and community debate. The intention is that the framework should guide description of practice and allow some comparison to be made. It will also allow students and practitioners to review what is happening in their own

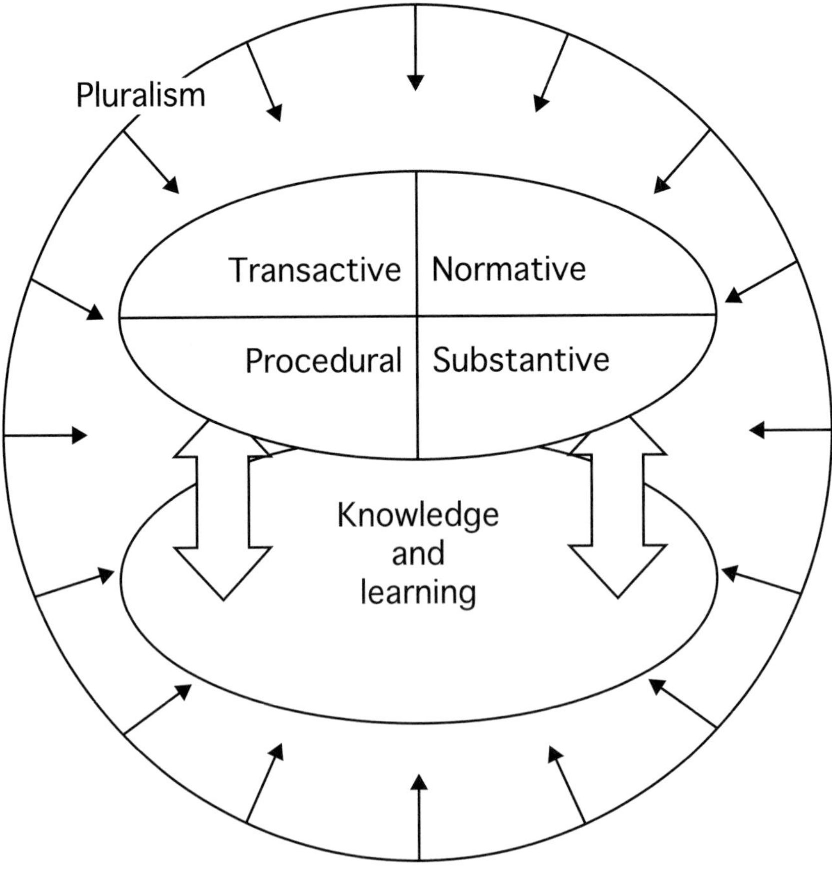

Figure 8.1

Typology of effectiveness

areas and compare this with the cases examined from Canada, England, Western Australia and South Africa in the following chapters (9 to 12).

Other comparative frameworks exist in impact assessment fields (see, for example, Sadler, 1996 detailed in Chapter 3). While different observers might place different emphasis or propose different criteria to those used here, this reflects the pluralistic nature of the field which we have already discussed. Exercising judgement on effectiveness is fundamental to any comparative review, and we propose that consideration of procedural, substantive, transactive and normative effectiveness, along with consideration of how pluralism, learning and knowledge in the field are addressed is essential. Following the insights provided in Chapters 1 and 2 we understand the key principles which need to be embedded in any sustainability assessment process, and that it needs to be theoretically underpinned. In Chapters 4 to 7, we recognised that issues of time horizons, geographical scales, legal systems and professional standards also all need to be considered. Below we discuss how each of these criteria contributes to any evaluation of effectiveness. In doing this, we draw on the literature which has, for some types of effectiveness, already identified a number of specific criteria. Where they exist, we present these criteria – not to develop a long and complicated checklist of effectiveness (which we consider time consuming and unhelpful), but to clarify the nature and significance of 'effectiveness' in specific settings. In this way we hope that what constitutes procedural, substantive, normative or transactive effectiveness, and what effective sustainability assessment will do in relation to pluralism, learning and knowledge will be clear. The chapter aims to offer a clearer understanding of what we think should be considered as effective sustainability assessment practice in any specific setting. It also offers some simple questions around this understanding to guide the evaluation of existing sustainability assessment practice in Chapters 9 to 12, and also what we consider are potential 'solutions' to problems of ineffectiveness (Chapters 13 to 16).

In the preface to this book we emphasised the importance of not taking literature for granted. In the context of this chapter this means taking into consideration that in developing the framework, we (the chapter authors) have at times had to adopt a particular framing in relation to the critical debates covered in Chapter 3 (for example, we have tried to avoid the significant reductionism that would be reflected through the development of overly comprehensive sets of detailed criteria, but are not holistic as we have developed a six-point typology – so our frame falls somewhere between the polar opposites in critical debate 4). This is unavoidable, but any reading of the evaluations in Chapters 9 to 12 must take this into account as the framework used will have influenced the conclusions reached in those chapters evaluating the effectiveness of practice. Applying the framework also relies on value-judgements and that means that the subsequent evaluations in Chapters 9 to 12 naturally would likely be disputed by others applying the same framework (the chapter authors have been selected based on their experience with practice in those jurisdictions; we consider their respective evaluations to be appropriately informed and valuable perspectives, but not the only perspectives). The questions we develop in our framework avoid the use of yes or no answers as this implies a kind of rationality and/or certainty which is not present; we prefer to use more open-ended questions for the evaluator to consider and exercise judgement against. We encourage the disputes that this approach will entail, but would argue that the evaluation questions cover a broader range of value-judgements than has previously been investigated. We would not argue that they are the final word on the matter and cannot be improved.

Procedural effectiveness

Published comparative reviews for project-level EIA and SEA (see, for example, Wood, 2003; Jones *et al.*, 2005) have been based largely on procedural criteria. Their focus is primarily on analysis of compliance with both regulatory stages and expectations of good practice. This can be seen in Table 8.1 which sets out all the criteria from Wood (2003) and the relevant (procedural) criteria from Jones *et al.* (2005, pp.41–42).

In most jurisdictions, the trigger for any formal (external) environmental assessment is determined by specific legislative and regulatory means. Administrative and regulatory procedures, therefore, are central to the way that an assessment is undertaken; they lay out the steps that must be followed. Inevitably, in any evaluation of effectiveness in these jurisdictions, procedural criteria will be abundant and easily identified by reviewing the relevant legislation or administrative procedures. Because they are relatively easy to measure, they risk overwhelming a more comprehensive consideration of effectiveness. However, we will simplify the evaluation of the procedural effectiveness of the sustainability assessment process and ask: Have appropriate processes been followed that reflect institutional and professional standards and procedures? On its own this will not tell us how sustainable outcomes are, but it is still an important and relevant question because if practice diverges from expectation set out in administrative and regulatory procedures it is still unlikely that it would lead to sustainable outcomes. In the case of less formal assessments, referred to in Chapter 7 as internal sustainability assessment, the question might be rephrased to consider how procedures adopted to undertake the assessment reflect and respond to concerns about institutional and professional standards and procedures.

Substantive effectiveness

Substantive effectiveness in the context of sustainability assessment means that more sustainable outcomes are achieved than likely would be the case in the

Table 8.1 Procedural criteria used to evaluate EIA and SEA systems

Environmental Impact Assessment (Wood, 2003)	Strategic Environmental Assessment (Jones *et al.*, 2005)
Is the EIA system based on clear and specific legal provision?	Are there clear legal provisions, defining broad objectives, standards and terms of reference, to undertake the SEA of land use plans?
Must the relevant environmental impacts of all significant actions be assessed?	Is there provision for the early integration of SEA and land use plan preparation?
Must evidence of the consideration, by the proponent, of the environmental impacts of reasonable alternative actions be demonstrated in the EIA process?	Does guidance relating specifically to the SEA of land use plans exist?
Must screening of actions for environmental significance take place?	Must the significant environmental effects of all land use plans be subjected to SEA?
Must scoping of the environmental impacts of actions take place and specific guidelines be produced?	Is the SEA of land use plans undertaken within a tiered system of environmental assessment?
Must EIA reports be publicly reviewed and the proponent respond to the points raised?	Is the concept of sustainable development integral to the SEA process?
Must the findings of the EIA report and the review be a central determinant of the decision on the action?	Does the SEA process provide for the consideration of reasonable alternatives, and must reasons for the choice of the selected alternatives be outlined?
Must monitoring of action impacts be undertaken and is it linked to the earlier stages of the EIA process?	Must screening of land use plans for environmental significance take place?
Must the mitigation of action impacts be considered at the various stages of the EIA process?	Are the boundaries of SEAs determined using scoping procedures?
Must consultation and participation take place prior to, and following, EIA report publication?	Are the policies within land use plans assessed against environmental criteria, and is the significance of the potential impacts evaluated?
Must the EIA system be monitored and, if necessary, be amended to incorporate feedback from experience?	Does the SEA process explicitly require consideration of secondary, synergistic or cumulative impacts?
Are the discernible environmental benefits of the EIA system believed to outweigh its financial costs and time requirements?	Are the SEA procedures and their main findings recorded in publicly available SEA reports?
Does the EIA system apply to significant programmes, plans and policies, as well as to projects?	Is the information included in SEA reports subjected to a transparent review process to check that it is sufficient to inform decision making?
	Do SEAs include monitoring strategies linked to the achievement of pre-defined objectives for land use plans?
	Does a mitigation strategy exist to promote environmental enhancement and the reduction of potentially negative environmental effects?
	Does consultation and public participation take place within the SEA process, and are the representations recorded and acted upon?

absence of a sustainability assessment process (e.g. such as traditional EIA or some other appraisal tool). In many respects substantive effectiveness lies at the heart of sustainability assessment. However, the reason we do not focus on it to the exclusion of the other types represented in Figure 8.1 is because it would be unachievable in isolation from the other types.

Theophilou *et al.* (2010) focussed on the effectiveness of SEA applied in a particular context (of European Union structural funds) and, drawing on a range of literature (Sadler, 1996; European Parliament and the Council of the European Union, 2001; Cashmore, 2004; Fischer and Gazzola, 2006; GRDP Project, 2006; Jay *et al.*, 2007) identified a range of substantive criteria. However, as the authors used only substantive and transactive criteria, there are overlaps with other types of effectiveness set out in Figure 8.1 (such as the evidence for learning which we consider separately under 'knowledge and learning'). Similarly, Jones *et al.* (2005) also detailed outcome criteria albeit with similar overlaps across the effectiveness types set out in Figure 8.1. Thérivel and Minas (2002) looked at the route to effective appraisal in the context of environmental and/or sustainability appraisal of UK development plans, and Prendergast *et al.* (2008) developed a process checklist for SEA for the Irish Environmental Protection Agency which specifies some substantive outcomes. The substantive effectiveness criteria drawn from these four sources are summarised in Table 8.2.

To evaluate the substantive effectiveness of sustainability assessment processes, we will ask: In what ways, and to what extent does sustainability assessment lead to changes in process, actions, or outcomes?

Table 8.2 Substantive criteria used to evaluate an SEA system (adapted from Thérivel and Minas, 2002; Jones *et al.*, 2005; Prendergast *et al.*, 2008; Theophilou *et al.*, 2010)

Has the SEA had any effect 'on the ground' in terms of improving the environmental quality of the area?
Has the SEA process informed decisions on the final version of the plan or programme?
Have the statutory consultation bodies had a fair opportunity to contribute and have their views and comments been taken on board?
Has the SEA had any discernible influence on the content of land use plans or the treatment of environmental issues during decision making?
Does any form of monitoring of the SEA process and outcomes take place?
Does the SEA help to ensure that development is within environmental limits?
Has the SEA process suggested sustainable new alternatives that were actively considered?
Are the mitigation measures proposed by the SEA commensurate with the type and scale of impacts of the plan?
Have the SEA's mitigation measures been incorporated into the plan?

Table 8.3 Transactive criteria used to evaluate an SEA system (adapted from Theophilou *et al.*, 2010)

Has the SEA been carried out within a reasonable time frame without undue delay?
Has carrying out the SEA entailed reasonable (as opposed to excessive) spending?
Has acquiring the requisite skills and personnel for the SEA constituted a big burden or were they easily accessible?
Were responsibilities clearly defined and allocated and tasks undertaken by the most appropriate subjects?

Transactive effectiveness

The practicalities of any assessment process dictate that it must be a cost-effective exercise (Finsterbusch, 1995) and evaluation on this basis is necessary. Baker and McLelland (2003, p.584) equate transactive effectiveness with 'proficiency' and based their understanding of the term on that used by Sadler (1996) in the International Effectiveness of Environmental Assessment Study. Theophilou *et al.* (2010) determined a number of specific criteria to measure transactive effectiveness associated with SEA application as set out in Table 8.3.

Evaluation against these criteria is inevitably subjective in that views will differ on what is a reasonable time frame and what is reasonable spending. However, for the purposes of evaluating sustainability assessment, we have encapsulated these criteria with the question: To what extent, and by whom, is the outcome of conducting sustainability assessment considered to be worth the time and cost involved?

Normative effectiveness

Normative decision making is generally conceptualised in terms of a distinction between personal and social norms (Klöckner and Matthies, 2004) with personal norms being based on the moral viewpoints of the individual and social norms being the perceived expectations of others (i.e. society). Normative principles are commonly at play in all forms of assessment (and are likely to affect the evaluation of sustainability assessment against each of the criteria considered in this chapter). In the English Sustainability Appraisal system, a framework of desired sustainability objectives are derived which are, essentially, normative visions of the future. Of course, this distinction between individual and social norms will also influence the ways in which the criteria used for comparison are understood – claims of universal values should always be subject to some detailed debate and evidence as they are often assertions of an ideology rather than observations of agreed principles.

But what exactly are norms and what place do they have in sustainability assessment? Norms were (somewhat simplistically) defined in Chapter 3. However, in the context of sustainability, we are actually seeking a social sustainability norm (individual sustainability norms are dealt with through the pluralism criterion below) so that we can judge whether a particular sustainability assessment process is appropriately focussed from a normative perspective. Hartmuth *et al.* (2008) define a sustainability norm by considering the Brundtland definition of sustainable development (World Commission on Environment and Development, 1987) from the viewpoint of ethical justice. In doing so, they derive sustainability goals from which they derive sustainability rules.

In this book where we focus on an assessment process, we define the sustainability norm as being the imperatives of sustainability assessment set out by Gibson in Chapter 1.[1] These are derived from a combination of empirical and theoretical evidence and provide a yardstick for gauging normative effectiveness. Thus, to evaluate normative effectiveness, we ask the question: In what ways, and to what extent does the sustainability assessment satisfy the following imperatives:[2]

- reverse prevailing (unsustainable) trends?
- integrate all the key intertwined factors affecting sustainability?
- seek mutually reinforcing gains
- minimise trade-offs?
- respect contexts in which sustainability assessment takes place?
- is open and broadly engaging?

Pluralism

A number of authors have grappled with the theoretical underpinnings of impact assessment processes (see, for example, Lawrence, 1997; Bartlett and Kurian, 1999; Lawrence, 2000; Cashmore, 2004; Elling, 2009; Howitt, 2011) and some of this debate has been covered by Cashmore and Kørnøv in chapter 2 of this book. Two clear conclusions on which scholars seem agreed are:

1 Decision making is not rational and therefore a positivist theory of impact assessment whereby better information always leads to better decisions is flawed.
2 No single theory fully explains the influence and effectiveness of impact assessment processes.

Owens *et al.* (2004, p.1947) argue that there are too many contested frames for assessment to ever be 'a neutral or objective exercise' and this points to a fundamental difficulty – the problem is viewed differently by different actors, and so the goals of assessment are also contested. Lawrence concurs and refers to 'fundamentally different perspectives on [. . .] the appropriate role of EIA in decisionmaking' (Lawrence, 2003, p.xii), whilst O'Faircheallaigh (2009, p.107) focuses on the closely related process of Social Impact Assessment and argues that 'there is no consensus on what SIA is or on what are its purposes, and so no single definition of "effective SIA" is possible'.

Since it will always be possible for those with a specific framing to contest the outcomes of sustainability assessment with those who favour other framings, Bond *et al.* (2011, p.1161) argue:

> rather than adopt a specific framing for effectiveness, or for different interpretations of sustainability, there is a need to accommodate the different framings of stakeholders and, therefore, to lead to a process which is more likely to be seen as effective. This requires a more pluralistic approach to sustainability assessment as it accepts the presence of multiple value systems and the need to include all perspectives.

One means of integrating pluralism into sustainability assessment practice is through greater engagement with stakeholders and members of the diverse public. On a basic level, this means achieving higher levels of participation (there are many different levels of participation recognised, see for example, Arnstein, 1969) through the sustainability assessment process. In Europe, the need for greater involvement of the public in environmental decision making was recognised through the adoption of the United Nations Economic Commission for Europe's *Convention on Access to Information, Public Participation in Decision Making and Access to Justice in Environmental Matters* (United Nations Economic Commission for Europe, 1998) (known as the Århus Convention). Superficially, this has led to the amendment to the EIA Directive (European Parliament and the Council of the European Union, 2003) to require 'early' and 'effective' participation in the EIA process, although neither of these terms has been defined (Hartley and Wood, 2005). Palerm (1999), assessed the Århus Convention against principles derived from Habermas' theory of communicative action (which proposes a normative ideal for free speech to counter the tendency for those possessing power, e.g. decision makers, to impose their own framings on decisions) and found that the Convention fell short of the principles in four respects:

1. failure to ensure participation for cognitively and lingu[istic]ally non-competent actors
2. failure to ensure a two-way communication process
3. failure to ensure normative and subjective claims are adequately recognised
4. failure to establish conflict management procedures.

As well as the Århus Convention falling short of principles we argue are synonymous with accommodating pluralism, there is further evidence that the implementation of the provisions of the Convention in individual nation states is also variable (Hartley and Wood, 2005). The position we adopt here is that accommodating pluralism is the only way to deal with competing frames in sustainability assessment. Accommodating pluralism requires not simply the process of engagement with all those parties likely to be affected as well as other expert stakeholders at particular process stages, it assumes that different framings of problems and goals of assessment are valid and tries to ensure the process incorporates these frames. In particular, accommodating pluralism needs to be cognisant of the potential views of future generations to ensure intergenerational equity is properly ensured (albeit we can only make assumptions about the views of people not yet born).

In the context of sustainability assessment we will ask: how, and to what extent are affected and concerned parties accommodated into and satisfied by the sustainability assessment process?

Knowledge and learning

The management of knowledge and learning by all stakeholders, but especially regulators of impact assessment processes, is one important mechanism for improving practice over time (e.g. Sánchez and Morrison-Saunders, 2011). The use of knowledge in assessment as a means of facilitating better policies through a rational decision-making process is considered by Hertin *et al.* (2009) to be inherently flawed because of the political realities of policy making. Bond *et al.* (2010) emphasise the importance of knowledge in EIA, but suggest that informal knowledge (of participants in the EIA process) is critical in reaching a common understanding of sustainability goals. This acknowledges a multiplicity of views of sustainability assessment goals and suggests that this pluralism produces quite different sorts of knowledge and understanding amongst different stakeholders. As Lakoff (2004) notes, this produces different sorts of 'common sense' regarding what is 'good' sustainability assessment practice.

Jha-Thakur *et al.* (2009) considered the role of learning in determining the effectiveness of SEA and placed knowledge in the context of levels of learning drawn from the work of Bloom (1956). These levels move from knowledge, through comprehension, application, analysis, synthesis and evaluation. They refer to 'single-loop learning' and 'double-loop learning', the former of which corresponds to 'instrumental learning', and the latter to 'conceptual learning' in Nilsson's (2005, p.209) examination of environmental policy integration. Nilsson defines 'instrumental learning' in the context of policy changes as being learning that leads to the modification of policy to better achieve objectives whereas 'conceptual learning' refers to changes in beliefs (therefore fundamentally altering perspectives on policy and ways of achieving it). However, he offers up an additional classification of 'political learning' which he defines as 'the improvement of strategies, through argument and symbolic action' (indicating that a real shift in beliefs has not taken place).

Parallels can be drawn from project EIA and Social Impact Assessment, where there has long been a call for follow-up activities to improve the process and some progress towards that end (see, for example, Arts *et al.*, 2001; Gagnon, 2003; Morrison-Saunders *et al.*, 2003). The argument is that without follow-up (involving monitoring post-project implementation), the efficacy of prediction techniques used in the assessment remains uncertain. Thus, for EIA, the absence of follow-up means that the opportunity for instrumental learning has been missed.

In the context of sustainability assessment, the suggestion is that the conduct of the process itself not only contributes knowledge which can be used in different ways, it can also facilitate different levels of learning. Achieving instrumental learning is a benefit in terms of achieving sustainable outcomes – but only in the specific context of an individual case. Achieving conceptual learning can achieve longer-lasting benefits in that mindsets of actors are changed to recognise that sustainability assessment practice itself needs modification in order to better deliver sustainable outcomes. The value of conducting sustainability assessment, whilst actively learning from the experience, was identified by Bond *et al.* (2011) who subsequently modified their sustainability assessment practice to accommodate the learning.

In the context of sustainability assessment, we will ask: how, and to what extent, does the sustainability assessment process facilitate instrumental and conceptual learning? We acknowledge that other forms of learning have been identified (e.g. political learning), but wish to focus on the actual learning which takes place rather than symbolic actions.

Conclusions

The importance of norms and plural values, by definition, means that there will be disagreement over the extent to which sustainability assessment succeeds. The framework provided in this chapter, however, offers a set of criteria which is not prescriptive in terms of what sustainability assessment should deliver or how it should be delivered. Rather it raises questions that allow the evaluation of sustainability assessment systems and practices and facilitate comparison across times and places.

The following four chapters apply the criteria derived in this chapter to sustainability assessment as practised in England, Western Australia, Canada and South Africa. These are very different contexts and it should be borne in mind that these cases are as much a test of the criteria, as they are a test for the criteria. Care has been taken to ensure that the criteria are few in number and loosely defined, albeit within the constraints of the arguments made for their applicability. As such, the authors of the following four chapters can interpret the processes they evaluate based on their own context, and the answers to some of the questions are necessarily judgemental.

The criteria are summarised in Table 8.4.

Table 8.4 Framework for comparison of sustainability assessment processes

Framework criterion	Question asked
Procedural effectiveness	Have appropriate processes been followed that reflect institutional and professional standards and procedures?
Substantive effectiveness	In what ways, and to what extent, does sustainability assessment lead to changes in process, actions, or outcomes?
Transactive effectiveness	To what extent, and by whom, is the outcome of conducting sustainability assessment considered to be worth the time and cost involved?
Normative effectiveness	In what ways, and to what extent, does the sustainability assessment satisfy the following imperatives (set out by Gibson in Chapter 1): • reverse prevailing (unsustainable) trends? • integrate all the key intertwined factors affecting sustainability? • seek mutually reinforcing gains • minimise trade-offs? • respect contexts in which sustainability assessment takes place? • is open and broadly engaging?
Pluralism	How, and to what extent, are affected and concerned parties accommodated into and satisfied by the sustainability assessment process?
Knowledge and learning	How, and to what extent, does the sustainability assessment process facilitate instrumental and conceptual learning?

Notes

1 For individual proponents or businesses conducting their own internal sustainability assessments, the particular organisational culture or goals will largely determine the sustainability norms that apply.
2 Set out by Gibson in Chapter 1.

References

Arnstein SR (1969) 'A ladder of citizen participation', *Journal of the American Institute of Planners*, 35(4), 216–244.

Arts J, Caldwell P, Morrison-Saunders A (2001) 'EIA follow-up: good practice and future directions – findings from a workshop at the IAIA 2000 conference', *Impact Assessment and Project Appraisal*, 19(3), 175–185.

Baker DC and McLelland JN (2003) 'Evaluating the effectiveness of British Columbia's environmental assessment process for first nations' participation in mining development', *Environmental Impact Assessment Review*, 23(5), 581–603.

Bartlett RV and Kurian PA (1999) 'The theory of environmental impact assessment: implicit models of policy making', *Policy & Politics*, 27(4), 415–433.

Bloom B (ed.) (1956) *Taxonomy of Educational Objectives: The Classification of Educational Goals. Handbook 1: Cognitive Domain* (London: Longman Group Ltd).

Bond A, Dockerty T, Lovett A, Riche AB, Haughton AJ, Bohan DA, Sage RB, Shield IF, Finch JW, Turner MM, Karp A (2011) 'Learning how to deal with values, frames and governance in sustainability appraisal', *Regional Studies*, 45(8), 1157–1170

Bond AJ, Viegas CV, Coelho de Souza Reinisch Coelho C, Selig PM (2010) 'Informal knowledge processes: the underpinning for sustainability outcomes in EIA?' *Journal of Cleaner Production*, 18(1), 6–13.

Boothroyd P, Knight N, Eberle M, Kawaguchi J, Gagnon C (1995) 'The need for retrospective impact assessment: the megaprojects example', *Impact Assessment*, 13(3), 253–271.

Cashmore M (2004) 'The role of science in environmental impact assessment: process and procedure versus purpose in the development of theory', *Environmental Impact Assessment Review*, 24(4), 403–426.

Elling B (2009) 'Rationality and effectiveness: does EIA/SEA treat them as synonyms?' *Impact Assessment and Project Appraisal*, 27(2), 121–131.

European Parliament and the Council of the European Union (2001) 'Directive 2001/42/EC of the European Parliament and of the Council of 27 June 2001 on the assessment of the effects of certain plans and programmes on the environment', *Official Journal of the European Communities*, L197, 30–37.

European Parliament and the Council of the European Union (2003) 'Directive 2003/35/EC of the European Parliament and of the Council of 26 May 2003 providing for public participation in respect of the drawing up of certain plans and programmes relating to the environment and amending with regard to public participation and access to justice Council Directives 85/337/EEC and 96/61/EC', *Official Journal of the European Communities*, L156, 17–24.

Finsterbusch K (1995) 'In praise of SIA – a personal review of the field of social impact assessment: feasibility, justification, history, methods, issues', *Impact Assessment*, 13(3), 229–252.

Fischer TB and Gazzola P (2006) 'SEA effectiveness criteria – equally valid in all countries? The case of Italy', *Environmental Impact Assessment Review*, 26(4), 396–409.

Gagnon C (2003) 'Methodology of social impact follow-up modeling'. In: Rasmussen RO and Koroleva NE (eds) *Social and Environmental Impacts in the North: Methods in Evaluation of Socio-economic and Environmental Consequences of Mining and Energy Production in the Arctic and Sub-Arctic* (Dordrecht, Boston & London: Kluwer Academic) 479–489.

GRDP Project (2006) *Handbook on SEA for Cohesion Policy 2007–2013* [Online] Available at: <http://ec.europa.eu/regional_policy/sources/docoffic/working/doc/sea_handbook_final_foreword.pdf> (accessed 8 February 2008).

Hartley N and Wood C (2005) 'Public participation in environmental impact assessment – implementing the Aarhus Convention', *Environmental Impact Assessment Review*, 25(4), 319–340.

Hartmuth G, Huber K, Rink D (2008) 'Operationalization and contextualization of sustainability at the local level', *Sustainable Development*, 16(4), 261–270.

Hertin J, Turnpenny J, Jordan A, Nilsson M, Russel D, Nykvist B (2009) 'Rationalising the policy mess? Ex ante assessment and the utilisation of knowledge in the policy process', *Environment and Planning A*, 41(5), 1185–1200.

Howitt R (2011) 'Theoretical foundations'. In: Vanclay F and Esteves AM (eds) *New Directions in Social Impact Assessment: Conceptual and Methodological Advances* (Cheltenham: Edward Elgar), 87–95.

Jay S, Jones C, Slinn P, Wood C (2007) 'Environmental impact assessment: retrospect and prospect', *Environmental Impact Assessment Review*, 27(4), 287–300.

Jha-Thakur U, Gazzola P, Peel D, Fischer TB, Kidd S (2009) 'Effectiveness of strategic environmental assessment – the significance of learning', *Impact Assessment and Project Appraisal*, 27(2), 133–144.

Jones CE, Baker M, Carter J, Jay S, Short M, Wood C (eds) (2005), *Strategic Environmental Assessment and Land Use Planning: An International Evaluation* (London: Earthscan Publications Ltd).

Klöckner CA and Matthies E (2004) 'How habits interfere with norm-directed behaviour: a normative decision-making model for travel mode choice', *Journal of Environmental Psychology*, 24(3), 319–327.

Lakoff G (2004) *Don't Think of an Elephant! Know Your Values and Frame the Debate* (Melbourne: Scribe).

Lawrence DP (1997) 'The need for EIA theory-building', *Environmental Impact Assessment Review*, 17, 79–107.

Lawrence DP (2000) 'Planning theories and environmental impact assessment', *Environmental Impact Assessment Review*, 20, 607–625.

Lawrence DP (2003) *Environmental Impact Assessment: Practical Solutions to Recurrent Problems* (New Jersey: Wiley-Interscience).

Morrison-Saunders A, Baker J, Arts J (2003) 'Lessons from practice: towards successful follow-up', *Impact Assessment and Project Appraisal*, 21(1), 43–56.

Nilsson M (2005) 'Learning frames and environmental policy integration: the case of Swedish energy policy', *Environment and Planning C*, 23(2), 207–226.

O'Faircheallaigh C (2009) 'Effectiveness in social impact assessment: Aboriginal peoples and resource development in Australia', *Impact Assessment and Project Appraisal*, 27(2), 95–110.

Owens S, Rayner T, Bina O (2004) 'New agendas for appraisal: reflections on theory, practice, and research', *Environment and Planning A*, 36(11), 1943–1959.

Palerm JR (1999) 'Public participation in environmental decision making: examining the Aarhus Convention', *Journal of Environmental Assessment Policy and Management*, 1(2), 229–244.

Prendergast T, Donnelly A, d'Auria L, Desmond M, O'Mahony T, Devlin R, O'Driscoll D, Cronin A, Delalieux S, Owens M (2008) *Strategic Environmental Assessment (SEA) Process Checklist: Consultation Draft 18th January 2008*. [Online] Available at: <http://www.epa.ie/downloads/advice/ea/SEA%20Process%20Checklist.pdf> (accessed 19 July 2011).

Sadler B (1996) *International Study of the Effectiveness of Environmental Assessment Final Report – Environmental Assessment in a Changing World: Evaluating Practice to Improve Performance* (Ottawa: Minister of Supply and Services Canada).

Sánchez LE and Morrison-Saunders A (2011) 'Learning about knowledge management for improving environmental impact assessment in a government agency:

the Western Australian experience', *Journal of Environmental Management*, 92(9), 2260–2271.

Theophilou V, Bond A, Cashmore M (2010) 'Application of the SEA Directive to EU structural funds: perspectives on effectiveness', *Environmental Impact Assessment Review*, 30(2), 136–144.

Thérivel R, Minas P (2002) 'Ensuring effective sustainability appraisal', *Impact Assessment and Project Appraisal*, 20(2), 81–91.

United Nations Economic Commission for Europe (1998) *Convention on Access to Information, Public Participation in Decision Making and Access to Justice in Environmental Matters* (Geneva: United Nations Economic Commission for Europe, Committee on Environmental Policy).

Wood C (2003) *Environmental Impact Assessment: A Comparative Review* (Edinburgh: Prentice Hall).

World Commission on Environment and Development (1987) *Our Common Future* (Oxford: Oxford University Press).

9 Sustainability assessment in England

Riki Thérivel, Oxford Brookes University and Levett-Therivel Consultants

Introduction

Sustainability appraisals (the term used in England, rather than sustainability assessments) of local land use plans have been carried out in England since 1999, English land use planners are reasonably confident in carrying out sustainability appraisals, and sustainability appraisals are also increasingly being carried out for national government policies. It is less clear, however, just how much these appraisals are contributing to more sustainable plan-making.

Sustainability appraisal in England has evolved from two separate strands of assessment. The first strand began in 1991, with the publication of government guidance documents recommending environmental appraisal of (unspecified) policies and local-level development plans. The early appraisals were short and basic, typically checking that the plan was in accordance with governmental environmental and planning advice, checking whether the plan's objectives and policies were internally consistent, and assessing the plan policies' likely environmental impacts using a matrix with the plan policies on one axis and environmental components on the other axis.

In 1999, this approach was broadened into sustainability appraisal: 'Local authorities are expected to carry out a full environmental appraisal of their development plan . . . The same methodologies used for environmental appraisal can be developed to encompass economic and social issues' (DoE, 1999, s.4.16). In 2004, the UK Planning and Compulsory Purchase Act (UK Government, 2004, Article 19(5)) made sustainability appraisals mandatory for the main forms of spatial plans; in magnificently concise but non-specific language the local planning authority were required to '(a) carry out an appraisal of the sustainability of the proposals in each document; (b) prepare a report of the findings of the appraisal'.

The second strand in the development of English sustainability appraisals was the implementation of the European Strategic Environmental Assessment Directive (European Parliament and the Council of the European Union, 2001). The SEA Directive, which is heavily based on the Environmental Impact

Assessment Directive, sets detailed requirements for assessing the environmental impacts of certain plans and programmes, including the need to collect baseline data, consider alternatives and monitor the actual impacts of the plan. The Directive's definition of 'environment' is relatively broad, encompassing population, human health and 'material assets' in addition to more traditional environmental topics such as water and biodiversity. It also requires consideration of the 'interrelationship between the above factors', leaving considerable scope for also considering social and economic issues.

These two strands came together in 2004, when the SEA Directive and the Planning and Compulsory Purchase Act – and their partly overlapping requirements – were implemented virtually concurrently. In England, the SEA Directive was implemented through the Environmental Assessment of Plans and Programmes Regulations 2004, and supported by a government *Practical Guide to the SEA Directive* (ODPM *et al.*, 2005). Some English plans require SEA but not sustainability appraisal and vice versa, but both requirements apply to the main local (and formerly also regional) level land use plans, and these account for at least half of all sustainability appraisals and SEAs carried out in England. For these plans,[1] government prepared additional guidance (ODPM, 2005) on how the requirements of the SEA Directive could be incorporated within a broader sustainability appraisal prepared under the Act. Table 9.1 shows the sustainability appraisal stages recommended by this guidance, including the (limited) consultation requirements. The 2005 sustainability appraisal guidance has since been replaced by new guidance (PAS, 2010) that focuses on local level plans and incorporates emerging good practice, and that should be read jointly with the *Practical Guide*. The rest of this chapter will refer to this joint sustainability appraisal/SEA process as 'sustainability appraisal'.

Most sustainability appraisals carried out in England follow the structure of Table 9.1, both in the sequencing of the sustainability appraisal work and in the structure of the sustainability appraisal reports. So, for instance, a full range of social, economic and environmental policies and plans is analysed at Task A1, and the full social, environmental and economic baseline is described at Task A2. This approach allows sustainability considerations to be considered in an integrated manner rather than in separate silos.

Task A4, developing the sustainability appraisal framework, acts as the lynchpin of the assessment: the intersection of the social, economic and environmental components of sustainability appraisal, and also the intersection of the evidence collection and impact prediction stages. The sustainability appraisal framework reflects policy objectives and existing problems, and aims to provide a balance (or, if lucky, an integration) of social, economic and environmental issues. The sustainability appraisal framework is subsequently used at Stage B to assess the impacts of emerging alternatives, plan objectives, and plan policies. Box 9.1 shows a typical example of a simple sustainability appraisal framework. Although some of the sustainability appraisal objectives

Table 9.1 Incorporating sustainability appraisal within the plan-making process (adapted from ODPM, 2005)

Plan-making stage	Sustainability appraisal stage and tasks	Resulting reports
Pre-production – Evidence gathering	**Stage A: Setting the context and objectives, establishing the baseline and deciding on the scope** A1. Identifying other relevant policies, plans and programmes, and sustainability objectives A2. Collecting baseline information A3. Identifying sustainability issues and problems A4. Developing the sustainability appraisal framework A5. Consulting on the scope of the sustainability appraisal	Scoping report – sent to statutory consultees for comments, and typically made available on the planning authority website
Production	**Stage B. Developing and refining options and assessing effects** This stage involves several rounds of appraisal of the emerging plan using the sustainability appraisal framework as an appraisal structure, and consideration of ways of mitigating adverse effects and maximising beneficial effects. Typical appraisal stages are plan objectives, plan issues and options, the draft plan and the submission plan **Stage C. Preparing the sustainability appraisal report** **Stage D. Consulting on the preferred plan option and sustainability appraisal report** D1. Public participation D2i. Appraising significant changes made to the plan in response to public participation	One or more appraisal reports, culminating in a formal sustainability appraisal report and non-technical summary which are made available for consultation to the statutory consultees and the public
Examination	D2ii. Appraising significant changes resulting from representations	Possibly additional appraisal report(s)
Adoption and monitoring	D3. Making decisions and providing information	'SEA statement'
	Stage E. Monitoring the significant effects of implementing the DPD E1. Finalising aims and methods for monitoring E2. Responding to adverse effects	Monitoring reports

are clearly social (5) or environmental (6) or economic (12), most are a combination of several of these.

The last few years have seen a trend towards 'topic based' sustainability appraisal reports which are structured more like traditional project environmental impact statements, with each chapter focusing on a different sustainability topic. This approach has less potential to lead to fully integrated sustainability appraisal than the task-based approach of Table 9.1, although

Box 9.1 Typical example of sustainability appraisal framework composed of sustainability appraisal objectives (Leicester City Council, 2008)

1 To ensure that the existing and future housing stock meets the housing needs of the City
2 To improve health and reduce health inequalities
3 To provide better opportunities for people to value and enjoy the City's heritage and participate in cultural and recreational activities
4 To improve community safety, reduce crime and the fear of crime
5 To support diversity, tackle inequality, and support the development and growth of social capital across the communities of the City
6 To increase bio-diversity levels across the City
7 To protect, enhance and manage the rich diversity of the natural, cultural and built environmental and archaeological assets of the City
8 To manage prudently the natural resources of the City including water, air quality, soils and the minimising of flood risk through sustainable forms of development
9 To reduce the potential impact of climate change by minimising energy usage and to develop the City's renewable energy resource, reducing dependency on non-renewable resources
10 To make efficient use of existing transport infrastructure, help reduce the need to travel by car, improve accessibility to jobs and services for all and to ensure that all journeys are undertaken by the most sustainable mode available
11 To minimise waste and to increase the re-use and recycling of waste materials
12 To create high quality employment opportunities and develop a strong diverse and stable local economy
13 To raise the levels of educational achievement and develop a strong culture of enterprise and innovation
14 To reduce levels of deprivation

it seems to be generally preferred by consultees and gives a better indication of a plan's cumulative impacts on individual receptors than the task-based approach.

By now, hundreds of plans have been subject to sustainability appraisal, and many authorities have gone through more than one cycle of sustainability appraisal. Some of the sustainability appraisals are carried out by consultants, some by the planners themselves, some by officials within the plan-making authority who did not write the plan (for instance a sustainability officer), and some by combinations of these. National government departments have also started to carry out 'appraisals of sustainability' of some of their main policies, notably the emerging National Policy Statements for energy, ports, road and rail, waste water, hazardous waste, water supply and aviation; the proposed high-speed rail link between London and the Midlands/North of England; and the 'eco towns' programme. These appraisals use a similar approach to that used for local land use plans.

Procedural effectiveness

Although the Planning and Compulsory Purchase Act only requires the preparation of a sustainability appraisal report, the SEA Directive lists what such a report must include (minus some social and economic components), and most English sustainability appraisal reports are at least broadly compliant with these requirements. A quality review of 117 sustainability appraisal reports carried out in early 2008 (Fischer, 2010) found that almost three-quarters of the reports were of satisfactory quality. However none was without minor omissions, and 27 per cent were of unsatisfactory quality. The plan description and environmental baseline were generally adequate, as was the report presentation. More problematic were the consideration of options, identification and evaluation of impact significance, sustainability appraisal recommendations and monitoring. Particular shortcomings included:

- Poor explanation of how the plan sets the framework for other activities, and what issues are addressed elsewhere.
- Insufficient consideration of options, with limited description of how options were identified and evaluated, and the consideration of 'pseudo alternatives' (for instance plan v. no plan even where the development of a plan was clearly required).
- Unclear explanation of how public participation affected the plan and sustainability appraisal.
- Insufficient explanation of what changes were made to the plan as a result of the sustainability appraisal.
- Poor explanation of uncertainties and other difficulties, and insufficient monitoring proposals.
- Poor links between the generation of alternatives and the collection of supporting baseline data (Fischer, 2010).

The quality of the sustainability appraisal reports was found to decrease as the plans evolved: the scoping and issues/options sustainability appraisal reports had better scores, but almost half of final stage sustainability appraisals (44 per cent) had some serious omission (Fischer, 2010).

Some of these problems may be improving as a result of legal challenges and threats of legal challenges. The first legal challenges have mostly related to whether the sustainability appraisals have considered 'reasonable alternatives'. The sustainability appraisal for the East of England Regional Spatial Strategy was successfully challenged on its the lack of consideration of alternatives to a late inclusion of Greenfield housing sites. The South West and South East Regional Spatial Strategies were threatened with similar challenges, were partly withdrawn with a view to being rewritten, but were abolished as part of the government removal of the regional planning level before the conclusion of this legal saga. The National Policy Statements on energy were threatened with a legal challenge on the basis that their sustainability appraisals did not consider reasonable alternatives, and that the sustainability appraisals should have considered the impacts of the projects resulting from the policy statements, rather than what difference the policy statements would make compared to current policy. The policy statements were subsequently withdrawn and rewritten. A local level sustainability appraisal was successfully challenged because new alternatives were not considered even though the plan changed materially over time (with much larger housing numbers). These challenges are making other planners very aware of the need for the sustainability appraisal to be legally compliant so as to avoid the expense, stress and uncertainty caused by a legal challenge.

A recent government-funded study on sustainability appraisal effectiveness and efficiency (CLG, 2010) suggests a range of improvements to the sustainability appraisal process. Procedural recommendations include:

- Plan-making should generate well thought out and clearly articulated alternatives.
- The scope of the appraisal should reflect the alternatives being considered. The sustainability appraisal scoping stage should determine what evidence is needed to help choose between alternatives and propose appropriate mitigation.

Substantive effectiveness

Most plans subject to sustainability appraisal in England are changed in some way as a result of the sustainability appraisal. Surveys of local planning authorities have been carried out both before and after the 2004 implementation of the SEA Directive and Planning and Compulsory Purchase Act (Thérivel and Minas, 2002; Thérivel and Walsh, 2006; Sherston, 2008; Thomas, 2008; Yamane, 2008). They show that about 80 per cent of core strategies (which are a key component of local land use plans) are currently changed as a result of sustainability appraisal, rising from about 70 per cent pre-2004 (Figure 9.1). Planners' explanations of why their plans were not changed as a result of sustainability appraisal are typically that: 1) the plan was already sustainable, and/or 2) the sustainability appraisal process was begun too late in the plan-making process. Those changes that are made are typically quite limited: changes to individual words in the plan ('we added mitigation, we fine-tuned the plan'), rather than a wholesale change in the plan's approach, or consideration of new significant alternatives (Figure 9.2).

This picture of sustainability appraisal as leading to quite limited changes is echoed by the procedural points about the early stages of sustainability appraisal being carried out better than the later stages. It suggests, frustratingly, that sustainability appraisals act as a compendium of baseline data and documentation of the decision-making process, but have limited effect in improving the plan. Some of this may be because the SEA Directive only

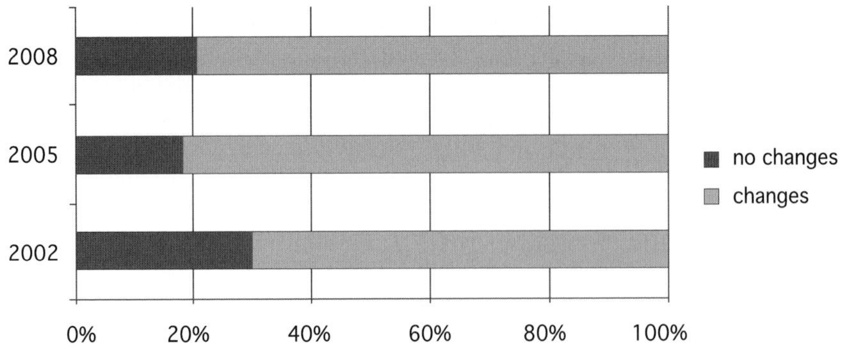

Figure 9.1

Proportion of plans changed as a result of sustainability appraisal/SEA, 2002, 2005, 2008 (based on Thérivel and Minas, 2002; Thérivel and Walsh, 2005; Sherston, 2008; Thomas, 2008; Yamane, 2008)

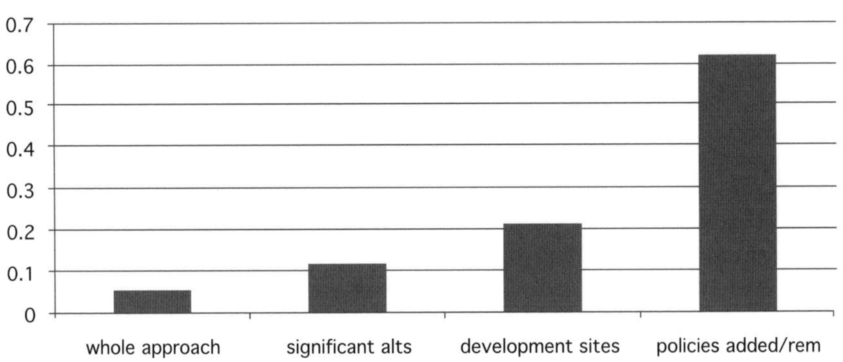

Figure 9.2

Type of changes, 2008 (based on Sherston, 2008; Thomas, 2008; Yamane, 2008)

requires appraisal findings to be 'taken into account' in decision making rather than establishing a stronger link. Some may be because the scoping report is often prepared in-house and the appraisal stage by outside consultants[2] (Fischer, 2010), raising concerns about whether the background evidence base of Stage A really informs the appraisal of Stage B, and whether the appraisal informs the plan-making processes.

The recent government study on sustainability appraisal effectiveness (CLG, 2010) concluded that, although very close integration of sustainability appraisal in the plan-making process helps to ensure that the sustainability appraisal is effective, it can lead to a loss of independence of appraisal, and it may not be clear afterwards how the sustainability appraisal has influenced the plan. To counter this problem, it recommended that an appraisal involving at least some people outside the planning team should be undertaken at key points in the plan-making process (notably at the alternatives and draft plan stages), and should provide plan-makers with explicit recommendations to which they can respond.

Transactive effectiveness

Due to the detailed requirements of the SEA Directive, and its broadening under sustainability appraisal/SEA to also include social and economic issues, sustainability appraisals do take quite a long time (and cost) to carry out: perhaps 40 to 60 person days for a standard sustainability appraisal/SEA, and longer for more complex plans.[3] That said, 68 per cent of English respondents to the 2008 survey believed that these costs would go down over time, due in part to increased familiarity with the appraisal process, and in part to the reduced time taken to update the scoping report after having to develop the first scoping report from scratch (Sherston, 2008; Thomas, 2008; Yamane, 2008).

Despite these costs in time, money and possible legal challenge, English planners still broadly felt in 2008 that the benefits of sustainability appraisal were worth the cost. Almost half of the English respondents to the 2008 survey of local authority planners – 46 per cent – agreed with the statement that sustainability appraisal is an effective use of time and resources, with 21 per cent being indifferent and (I think a remarkably low) 33 per cent disagreeing. Sustainability appraisal was felt to help to document the planning process, encourage 'planning as planning is meant to be', and help planners to resist market pressure. Although 41 per cent of respondents said that the sustainability appraisal process delayed their plan-making process (37 per cent slightly, 4 per cent significantly), more than twice as many said that other factors delayed the plan-making process (35 per cent slightly, 48 per cent significantly), with changing national and regional government policies and advice, elections and subsequent political changes, lack of resources, and lack of national guidance on plan-making being the most frequently cited reasons (Sherston, 2008; Thomas, 2008; Yamane, 2008).

Indications are that the recent economic recession and concomitant reductions in local government funding are causing local authorities to carry out more sustainability appraisals in-house, but with minimal or no additional resources. This would certainly increase the transactive effectiveness of sustainability appraisal, although it may well reduce the other forms of effectiveness.

In terms of transactive effectiveness, the government study (CLG, 2010) recommended that:

- Issues that are not likely to be significant should be scoped out of sustainability appraisals, as long as a clear explanation is given for doing so. Many English sustainability appraisals cover issues that are only tangentially related to the plan in question, possibly due to concerns over legal challenge. Tighter scoping would allow resources to be focused on the plan's most significant impacts, improving the sustainability appraisal's efficiency and clarity.

- Planners should integrate the evidence-gathering stages of the plan and sustainability appraisal. Planners already compile information on such issues as flooding, housing need, and access to facilities as part of the plan development process. They also already compile Annual Monitoring Reports on the implementation of their plans.[4] The sustainability appraisal process should piggyback on this information rather than starting from scratch, to reduce duplication and ensure that it focuses on issues of real concern to planners.
- Sustainability appraisals can act as umbrella documents for other forms of assessment, such as equalities impact assessment and health impact assessment, reducing duplication of analysis and report writing.

Normative effectiveness

Unfortunately, although English plans seem to be becoming *more* sustainable as a result of the sustainability appraisal, this does not mean that they *are* sustainable post-sustainability appraisal. In 2008, 45 sustainability appraisal reports for core strategies were analysed, to determine what they said about the sustainability of the core strategy (Thérivel *et al.*, 2009). Some of these plans were at the final submission stage, some at the earlier preferred option stage. The assumption was that a plan prepared with sustainable development as its core principle, and which had been subject to sustainability appraisal, would promote social, economic and environmental factors in a balanced way, ideally with positive effects across the board. The sustainability appraisal findings about the plans' sustainability were categorised according to 17 sustainability issues. Where the sustainability appraisal found that the plan had a very positive effect for that issue, a +2 was assigned; where it was very negative, a –2 was assigned; and impacts in between were assigned scores between +2 and –2.

Table 9.2 summarises the results of this analysis. On the whole, the sustainability appraisals concluded that the plans were positive in terms of social and economic issues, but somewhere between slightly positive and slightly negative for environmental issues. In many cases, the only negative impacts identified in the sustainability appraisal reports were environmental, or environmental factors were clearly affected significantly more negatively by the plan than were social or economic factors. More worrying, the environmental impacts of the submitted plans were significantly more negative than those of the preferred option documents, suggesting that plans become *less* sustainable as they get closer to completion (Thérivel *et al.*, 2009).

Table 9.2 Plan impacts on sustainability issues identified in the plan's sustainability appraisal report: finding of the analysis of 45 sustainability appraisal reports (Thérivel *et al.*, 2009)

Sustainability issue: sustainability appraisal (SA)/SEA topic		All SA/ SEAs	SA/SEA for. . .	
			preferred options document	Submitted Core Strategy
		n=45	n=28	n=17
Broadly social	Accessibility	1.27	1.20	1.38
	Crime	0.59	0.70	0.41
	Equity, inclusion	1.16	1.18	1.12
	Health	1.04	1.18	0.82
	Housing	1.23	1.21	1.26
	Average	**1.06**	**1.09**	**1.00**
Broadly environmental	Air	-0.21	0.04	-0.62
	Biodiversity	0.26	0.34	0.12
	Climate change, energy	0.09	0.38	-0.38
	Landscape, historical	0.67	0.63	0.74
	Resources	0.20	0.29	0.06
	Water	-0.04	0.11	-0.29
	Waste	-0.34	-0.21	-0.56
	Average	**0.09**	**0.23**	**-0.13**
Broadly economic	Economic growth, investment	1.18	1.21	1.12
	Employment	1.17	1.07	1.32
	Skills	0.68	0.68	0.69
	Average	**1.01**	**0.99**	**1.04**
	Flooding	-0.30	-0.12	-0.64
	Land use	1.04	1.11	0.94

Key: +2 very positive impact; 0 neutral impact; -2 very negative impact

If anything, the sustainability appraisals' conclusions about the plans' environmental impacts may have been over-positive. The sustainability appraisal objectives that comprise sustainability appraisal frameworks (Box 9.1) typically promote weak rather than strong sustainability, in that they promote directions of change ('to improve biodiversity') rather than requiring the achievement of stated sustainability appraisal targets. Furthermore, sustainability appraisals often test whether the plan is doing something about a topic – say encouraging the production of renewable energy or promoting non-car modes of travel – rather than whether the plan would actually improve it – say decrease greenhouse gas emissions. These sustainability appraisals essentially test whether the plan 'minimises' its own impacts rather than 'reduces' total environmental impacts. Many sustainability appraisals assume that all of the plan policies will be fully implemented as worded, rather than analysing what their likely implementation would be. For instance, plan policies on biodiversity were assumed to fully protect biodiversity, despite the fact that biodiversity had been declining in the past and other plan policies promoted significant new development. Some sustainability appraisals assumed that future careful project siting and design, or future project-level impact assessments, would neutralise the impacts of projects, with no clear justification for this assumption. In other words, the findings of Table 9.1 hide some assumptions that, if proven wrong in practice, may lead to more negative impacts (particularly environmental impacts) than predicted in the sustainability appraisals (Thérivel *et al.*, 2009).

Sustainability appraisals do seem to provide an environmental counterbalance to the prevailing socio-economic bias of plans. A 2008 questionnaire of all UK local planning authorities (Sherston, 2008; Thomas, 2008; Yamane, 2008) – a follow-on from previous surveys of 2002 and 2005, this time with a response rate of 32 per cent (151 out of 469) – asked whether planners felt that the plan-making and sustainability appraisal processes were biased in favour of the environment, economy or social issues, or evenly balanced between them. For the plan-making process, 42 per cent of respondents felt that it was biased towards the economy, 36 per cent that it was balanced, 15 per cent that it was more social, and 7 per cent more environmental. The main reason stated for this bias was the need to be consistent with national and regional economic and housing objectives. For the sustainability appraisal process, 53 per cent of the respondents felt that it was balanced, 39 per cent that it favoured the environment, 5 per cent social issues, and 3 per cent the economy. Interestingly, 70 per cent of respondents felt that the sustainability appraisal/SEA process changed the plan-making process to be more balanced, 24 per cent to be more environmental, and 6 per cent to be more social. Their additional comments suggested that they are broadly happy with sustainability appraisal's role as a balancing tool, for instance 'I am in agreement that the environment needs to take precedence because of greater long-term sustainability concerns' and 'The sustainability appraisal helps to balance the three strands of sustainable development'.

In terms of normative effectiveness, the government study (CLG, 2010) recommended that:

- To help avoid unrealistic assessment results, sustainability appraisals should assess plan impacts with reference to the baseline situation (*will the future baseline with the plan be better or worse than the future baseline without the plan?*), rather than with reference to sustainability appraisal objectives (*will the plan help to achieve social/environmental/economic objectives?*).
- The sustainability appraisal should consider the extent to which options and policies will be effectively delivered on the ground to help avoid unrealistic assessment results.

Pluralism

The SEA Directive requires, at minimum, consultation with environmental and other statutory authorities at the scoping and draft plan stages, and consultation with the public at the draft plan stage. In practice, English plans typically go through more than one round of public consultation (though the plans vary in terms of what stages are consulted on), and sustainability appraisal reports are typically prepared and made public as part of each of these plan consultations. Scoping reports and assessment reports are typically published on the Internet, but more active consultation such as workshops and public meetings on the sustainability appraisal are much less frequent.

The statutory authorities frequently comment on the sustainability appraisal reports, their comments are taken seriously by planning authorities, and subsequent sustainability appraisal reports often document these comments and how they were responded to. However, public involvement in English sustainability appraisal processes is negligible. Sustainability appraisal reports are large, technical documents, the strategic nature of the planning process is off-putting, and planners put little effort into engaging the public in the sustainability appraisal process (though they generally consult widely on the emerging plan). In the 2008 survey (Sherston, 2008; Thomas, 2008; Yamane, 2008), planners felt that sustainability appraisal makes the plan-making process more transparent, but that it does not generate significant public interest (Figure 9.3). The government study on sustainability appraisal effectiveness (CLG, 2010) recommended that public participation in sustainability appraisal could be improved through stakeholder events focused on plan options, and through the preparation and dissemination of separate, understandable non-technical summaries of sustainability appraisal reports.

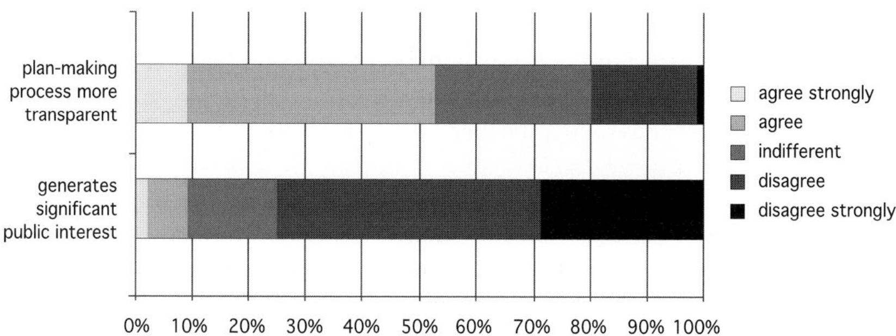

Figure 9.3

Pluralism effects of sustainability appraisal (based on Sherston, 2008; Thomas, 2008; Yamane, 2008)

Knowledge and learning

Although the picture described above is mixed, the 2008 survey of local authority planners suggests that the sustainability appraisal process does give planners clear knowledge and learning benefits in the form of greater awareness of sustainability issues, a better understanding of their plan, and ideas/ inspiration for the next round of plan-making (Figure 9.4) (Sherston, 2008; Thomas, 2008; Yamane, 2008). Additional benefits of sustainability appraisal, each listed by several survey respondents, included the creation of a sustainability officer post to deal with a variety of sustainability and environmental tasks in the planning department, greater emphasis on joint working between planners and external agencies during plan preparation, accumulation of background data to inform the wider plan-making process, and identification of the need for further information (e.g. about flooding or nature conservation).

England's community of sustainability appraisal researchers and practitioners is particularly active. Academic institutions carry out surveys and review sustainability appraisal report quality, consultancies voluntarily prepare sustainability appraisal guidance (including guidance on sustainability appraisal of neighbourhood plans (Levett-Therivel and URS/Scott Wilson, 2011)), and local planners participate in online forums on how to write and appraise plans. Much of this is a response to a vacuum left by a relative lack of government guidance and leadership in a very rapidly changing planning context. Planners are also trying to avoid legal challenge and improve the efficiency of all aspects of plan-making at a time of swinging cuts in their budgets.

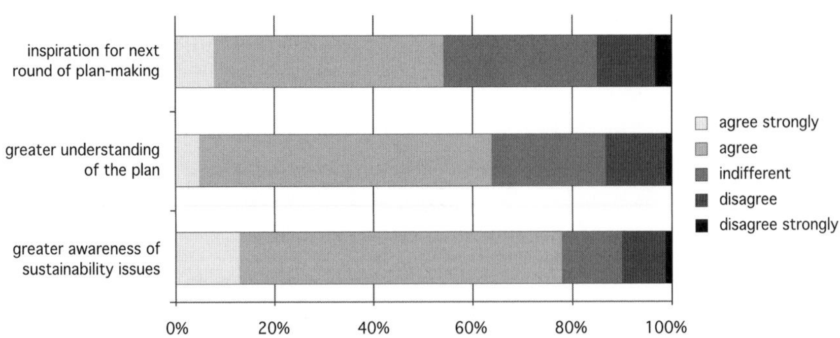

Figure 9.4

Knowledge and learning effects of sustainability appraisal (based on Sherston, 2008; Thomas, 2008; Yamane, 2008)

Table 9.3 English perspective on the effectiveness of sustainability appraisal

Framework criterion	Questions asked	English perspective
Procedural effectiveness	*Have appropriate processes been followed that reflect institutional and professional standards and procedures?*	English sustainability appraisals generally fulfil the legal requirements of the SEA Directive, but their consideration of plan alternatives is often poor, and other aspects could also be improved. The earlier stages of sustainability appraisal tend to be carried out better than the later stages.
Substantive effectiveness	*In what ways, and to what extent, does sustainability assessment lead to changes in process, actions, or outcomes?*	Sustainability appraisals generally lead to some changes in plans, but these tend to be minor: changes in individual words rather than in overall objectives or broad approaches.
Transactive effectiveness	*To what extent, and by whom, is the outcome of conducting sustainability assessment considered to be worth the time and cost involved?*	English sustainability appraisals are quite expensive, but this reflects the detailed and demanding requirements of the SEA Directive. Sustainability appraisals are expected to become cheaper as planners become more familiar with the process, and as they increase their efficiency in response to reduced funding as a result of the economic recession.
Normative effectiveness	*In what ways, and to what extent, does the sustainability assessment satisfy the listed normative imperatives?*	Sustainability appraisal in England is weak in its achievement of normative principles. Although sustainability appraisals help to 'rebalance' plans from a social and economic bias to a *more* overall sustainable position, and help to ensure that social, economic and environmental factors are considered in a *more* integrated fashion, they do not ensure that plans are fully sustainable or integrated. They generally do not help to achieve mutually reinforcing gains or minimize trade-offs: where difficult decisions need to be made, social and economic factors still consistently 'win' over environmental factors. Sustainability appraisals still often feel quite generic, although increased use of mapping and a move towards assessing plan impacts with reference to the baseline situation rather than sustainability appraisal objectives (which themselves are often generic) is helping to make sustainability appraisals more adapted to the particulars of their contexts. Sustainability appraisal reports are typically (only) made available on the Internet: this limited access, and the technical nature of sustainability appraisal

		does not lead to active public participation in the sustainability appraisal process. However the English plan-making process itself is typically open and broadly engaging.
Pluralism	*How, and to what extent, are affected and concerned parties accommodated into and satisfied by the sustainability assessment process?*	Expert environmental authorities must be consulted at several stages in the sustainability appraisal process, and their views are taken seriously. The general public's level of engagement in the sustainability appraisal process is low. However a series of high-profile legal challenges to sustainability appraisals is leading to sustainability appraisal being perceived, perhaps unhelpfully, as the 'soft underbelly' of planning, and so increasing the public's interest in sustainability appraisal as a tool by which to oppose plans.
Knowledge and learning	*How, and to what extent, does the sustainability assessment process facilitate instrumental and conceptual learning?*	English planners cite a range of indirect benefits of sustainability appraisal, including greater understanding of their plans, greater understanding of sustainability, and ideas for future rounds of planning. English academics and consultants are very active in researching sustainability appraisal practice and promoting new approaches to sustainability appraisal.

Conclusions

Table 9.3 summarises the level to which the English sustainability appraisal system achieves the framework criteria of Chapter 8. Strengths of the English sustainability appraisal system include a consistent and active attempt to promote sustainable development through the plan-making system; an active and often successful attempt to integrate the traditional environmental, social and economic silos; sustainability appraisal practice now spanning more than ten years, with many authorities having carried out several cycles of sustainability appraisals; the increasing application of sustainability appraisal to national as well as local level plans; and an active consultancy and research community that learns from, and disseminates information about, sustainability appraisals. Weaknesses include the lack of legal teeth of sustainability appraisals, whose findings must only be 'taken into account'; the weak form of sustainability promoted through the sustainability appraisal system; the resulting lack of full sustainability of the resulting plans; and the perception by many stakeholders that sustainability appraisal is an expensive, technocratic system that merely duplicates what planners already do anyway.

Recent major changes in the English planning system – notably the abolition of regional level plans and the promotion of 'localism' – have raised many questions about what future plans will look like, but the future of sustainability appraisal/SEA looks secure, in part because the SEA Directive is not going away. That said, seven years into the implementation of 'new style' sustainability appraisal, sustainability appraisal practice can clearly be strengthened and streamlined.

Notes

1 The UK planning system is currently in a state of enormous upheaval, with the expected abolition of regional level plans in spring/summer 2012 and the promotion of 'localism' in plan-making, hence the loose explanation here.
2 This may be done for reasons of impartiality – consultants are perceived as being less biased than the planners when the impacts of the plan are assessed. Planners may also start their planning process with high ambitions of doing the work in-house, but then outsource the sustainability appraisal when timescales get tight.
3 Specific information on this is difficult to get. This number is based on the limited responses to this question in the 2008 survey of local authority planners, and the author's consultancy experience.
4 In spring 2012, Government stopped requiring the production of Annual Monitoring Reports. It is unclear how SEA data will be monitored in the future.

References

CLG (Department for Communities and Local Government) (2010) *Towards a More Efficient and Effective Use of Strategic Environmental and Sustainability Appraisal in Spatial Planning: Final Report.* [Online] Available at: <http://www.communities.gov.uk/documents/planningandbuilding/pdf/1513010.pdf> (accessed 23 November 2011).

Department of the Environment (DoE) (1999) *Planning Policy Guidance Note 12: Development Plans* (London: HMSO).

European Parliament and the Council of the European Union (2001) 'Directive 2001/42/EC of the European Parliament and of the Council of 27 June 2001 on the assessment of the effects of certain plans and programmes on the environment', *Official Journal of the European Communities*, L197, 30–37.

Fischer T (2010) 'Reviewing the quality of strategic environmental assessment reports for English core strategies', *Environmental Impact Assessment Review*, 30(1), 62–69.

Leicester City Council (2008) *Core Strategy Sustainability Appraisal.* [Online] Available at: <http://www.leicester.gov.uk/EasySiteWeb/getresource.axd?AssetID=17483&type=full&servicetype=Attachment> (accessed 23 November 2011)

Levett-Therivel and URS/Scott Wilson (2011) *DIY SA: Sustainability Appraisal of Neighbourhood Plans.* [Online] Available at: <http://www.levett-therivel.co.uk/DIYSA.pdf> (accessed 14 February 2012).

Office of the Deputy Prime Minister (ODPM) (2005) *Sustainability Appraisal of Regional Spatial Strategies and Local Development Documents* (London: The Stationery Office).

ODPM, Scottish Executive, Welsh Assembly Government & DOE Northern Ireland (2005) *A Practical Guide to the SEA Directive.* [Online] Available at: <http://www.communities.gov.uk/documents/planningandbuilding/pdf/practicalguidesea.pdf> (accessed 23 November 2011).

Planning Advisory Service (PAS) (2010) *Sustainability Appraisal: Advice Note.* [Online] Available at: <http://www.pas.gov.uk/pas/aio/627078> (accessed 23 November 2011).

Sherston T (2008) *The Effectiveness of Strategic Environmental Assessment as a Helpful Development Plan Making Tool.* MSc dissertation (Oxford: Oxford Brookes University).

Thérivel R and Minas P (2002) 'Ensuring effective SEA in a changing context', *Impact Assessment and Project Appraisal,* 29(2), 81–91

Thérivel R and Walsh F (2006) 'The strategic environmental assessment directive in the UK: 1 year onwards', *Environmental Impact Assessment Review,* 26(7), 663–675.

Thérivel R, Christian G, Craig C, Grinham R, Mackins D, Smith J, Sneller T, Turner R, Walker D, Yamane M (2009) 'Sustainability-focused impact assessment: English experiences', *Impact Assessment and Project Appraisal,* 27(2), 155–168.

Thomas P (2008) *Four Years on from the Implementation of the SEA Directive.* MSc dissertation (Oxford: Oxford Brookes University).

UK Government (2004) *The Planning and Compulsory Purchase Act 2004.* [Online] Available at: <http://www.legislation.gov.uk/ukpga/2004/5/pdfs/ukpga_20040005_en.pdf> (accessed 24 November 2011).

Yamane M (2008) *Achieving Sustainability of Local Plan through SEA/SA.* MSc dissertation (Oxford: Oxford Brookes University).

10 Learning by doing: sustainability assessment in Western Australia

Angus Morrison-Saunders, Murdoch University and North-West University
Jenny Pope, Integral Sustainability and North West University

Introduction

Western Australia covers one-third of the Australian continent, is home to only slightly more than 10 per cent of the national population, and accounts for around 44 per cent of Australia's exports (DFAT, 2010). Its economic strength derives from the exploitation of the state's rich mineral resources, which include crude oil, natural gas, iron ore, gold, nickel, copper and other metals (DFAT, 2010). Western Australia also has a long and strong tradition of project-based environmental impact assessment (EIA) to which these major extractive projects are subject. For example, in his comparative review of EIA performance for around a dozen jurisdictions worldwide, Wood (1994, p.333) stated that: 'Widely perceived as a comprehensive and effective EIA system, Western Australia's EIA process is of particularly comparative interest'. One strength singled out by Wood (1994) is the independent Environmental Protection Authority (EPA) which administers and reports on EIA to the Minister for Environment.

However the scope of EIA in Western Australia is limited in legislation to mainly consideration of biophysical impacts (Bache *et al.*, 1996). The *Environmental Protection Act* 1986 (*EPAct*), under which EIA in Western Australia occurs, contains some sustainability provisions; these were added as s4A in the 2003 amendments to the Act, a time when the state government was actively pursuing sustainability assessment initiatives as noted in Chapter 7. Specifically s4A of the *EPAct* specifies that the object of the Act is to protect the environment of the state, having regard to the precautionary principle;

intergenerational equity; intragenerational equity; the principle of the conservation of biological diversity and ecological integrity; principles relating to improved valuation, pricing and incentive mechanisms; and the principle of waste minimisation. While the EPA does give some consideration to these principles in its application of EIA, ultimately it has not substantially deviated from its traditional focus on biophysical considerations.

The evolution of sustainability assessment in Western Australia has been characterised by a willingness on the parts of government, proponents and the community to experiment and to adopt a 'learning by doing' approach to this emerging decision-aiding tool, underpinned by a commitment to generating better outcomes from development for the community as a whole, as well as to 'make a case' for development projects. Early sustainability assessment processes were led by government and integrated with the formal project assessment and approval processes, including EIA, and therefore were examples of *external* sustainability assessment (see Chapter 7). Increasingly, however, proponents ranging from major corporations to small local governments have embraced and experimented with *internal* forms of sustainability assessment that guide their internal planning and decision-making processes. In some cases these processes are conducted in the early stages of a project that is subsequently subject to statutory EIA, but in others, particularly at more strategic levels of planning, it is undertaken purely for reasons of good governance.

In this chapter we describe some diverse Western Australian case studies of sustainability assessment, some of which were introduced in Chapter 7, highlighting innovative approaches to sustainability assessment as they have emerged in Western Australia in the absence of any statutory mandate or formal process. We draw primarily upon our experiences as practitioners and researchers in conducting this analysis. We conclude by presenting our findings within the framework for comparing and evaluating sustainability assessment practice established in Chapter 8.

The evolution of sustainability assessment in Western Australia

Although a number of organisations had been making efforts to incorporate sustainability thinking and planning and decision making for some time, the term 'sustainability assessment' came into common use in Western Australia in 2002 with the publication of the draft Western Australian State Sustainability Strategy, followed by the final Strategy the following year (Government of Western Australia, 2002; 2003). One early example of sustainability assessment was the Perth's Water Futures study conducted by the then Water Authority of Western Australia (now the Water Corporation) in 1995 which utilised a multi-criteria analysis (MCA) approach to select between water supply options. Although the terminology was not used, the criteria reflected sustainability considerations (Water Authority of Western Australia, 1995).

The 2003 State Sustainability Strategy included commitments that government would undertake sustainability assessments of complex and strategic projects, and that government agencies would apply sustainability assessment in internal decision making. While the latter commitment was never fully implemented, the period 2002–2005 saw two significant proposals subject to external regulatory sustainability assessment: the Gorgon Gas Development on Barrow Island (hereafter 'Gorgon') and the South West Yarragadee Water Supply Development (hereafter 'SWY'). A change of premier in Western Australia in January 2006 saw the government-led sustainability agenda quietly disappear, taking any active promotion of this form of sustainability assessment with it.

At this point the centre of gravity of sustainability assessment practice shifted from government to proponent activities. In the hands of a range of proactive proponents, including some from the resources sector but also public infrastructure providers and local governments, sustainability assessment practice has continued to evolve in different forms, arguably benefiting from the common language and understanding that emerged from the government-led processes. A body of sustainability assessment practitioners has emerged, and although the group is more disparate than it probably was during the government's period of experimentation with sustainability decision making, it is alive and well, as evidenced by robust attendance at two Sustainability Assessment Symposia convened in Perth in 2008 and 2010 (the proceedings of which are available at www.integral-sustainability.net).

The practice of sustainability assessment in Western Australia

In this section we present some examples and case studies of sustainability assessment from Western Australia. We commence by focusing in some detail on the two government-led sustainability assessment processes, before sketching the contours of current proponent-led sustainability assessment practice.

External regulator-led sustainability assessment

The Gorgon and SWY assessments were conducted by government, with the co-operation and collaboration of the relevant proponents, in the absence of any legal frameworks and in accordance with an active policy of 'learning by doing'. These case studies have been examined in some detail (DoIR, 2004; Pope *et al.*, 2005; Newman, 2006; Pope and Grace, 2006; Pope, 2007). They have also been introduced in Chapter 7 of this book, and their salient characteristics are summarised below.

Gorgon gas development on Barrow Island

The Gorgon assessment, conducted in 2002–2003 by the Government of Western Australia (Pope *et al.*, 2005, Pope and Grace, 2006), represented the first example of a sustainability assessment (but actually referred to as an 'integrated strategic assessment', acknowledging that it represented a step forward but perhaps not one that truly deserved the sustainability assessment label) undertaken as part of a formal project assessment and approvals process in Western Australia. The purpose of the process was to support the decision as to whether or not access would be granted to Barrow Island for the purposes of the proposed development.

The proposed development was unlikely to be found to be acceptable under the statutory (biophysically oriented) EIA process due to the potential for significant environmental impacts on Barrow Island, which had been a Class A Nature Reserve (i.e. the highest level of conservation protection status possible in Western Australia, requiring approval from both houses of Parliament to be amended) since 1910 and which has unique and internationally significant conservation values. In line with commitments made to introduce sustainability assessment processes for complex and strategic projects (Government of Western Australia, 2002; Independent Review Committee, 2002) it was decided that a new, integrated assessment process would be trialled for this highly controversial proposal. The intention was that a sustainability oriented assessment process would permit a more thorough and transparent examination of the strategic, social and economic, as well as the environmental implications of the proposal, and form a more appropriate basis for decision making.

The environmental assessment process was modelled on EIA as conducted in Western Australia, with the environmental assessment by the EPA mirrored by a non-statutory assessment of the strategic, social and economic implications of the proposal which was undertaken by an Expert Panel of consultants appointed by the Western Australian Department of Industry and Resources (DoIR). A scoping document was prepared identifying relevant strategic, environmental, social and economic issues; the proponent was required to prepare a draft Environmental, Social and Economic (ESE) Review which was then released for public comment; a final ESE Review was prepared, taking into consideration comments made; and then the EPA and the Expert Panel provided their advice to the Western Australian Cabinet, charged with making the decision. The Conservation Commission of Western Australia, the body in which the conservations estate is vested, also provided advice to government on the conservation implications of the proposal.

One of the key limitations of the Gorgon assessment was the extremely minimal consideration given to alternative locations for the development. The proponent, ChevronTexaco and its joint venture partners, had announced its intention to develop the Gorgon gas fields and argued that Barrow Island was the only commercially viable location for the necessary processing and shipping facilities, and this was the premise for the assessment process that followed. The proponent's own site selection analysis was reviewed by the Expert Panel, but under a confidentiality agreement to protect commercially sensitive data. A peer review of the proponent's MCA methodology applied to the site selection process found it to be flawed but no comprehensive, transparent assessment of alternative locations on the mainland was undertaken. Hence the decision faced by Cabinet was either to grant access to Barrow Island or not. This decision would clearly involve significant trade-offs whichever way the decision went, and the situation was exacerbated by a lack of clear sustainability objectives or criteria upon which to base the assessment (Pope *et al.*, 2005).

It was unsurprising to most commentators that the EPA and the Conservation Commission recommended to Cabinet that the development should not be allowed to proceed on environmental grounds, particularly the risk of the introduction of invasive animal and plant species into Barrow Island's sensitive ecosystem (EPA, 2003; Conservation Commission of Western Australia, 2003), while the Expert Panel recommended for the proposal on the grounds of positive socio-economic and state strategic benefits (Allen Consulting Group, 2003). Cabinet ultimately came down in favour of proceeding with development. Following the initial in-principle approval by Cabinet of the proposal, the proponent was required to submit more detailed development proposals as part of the formal EIA process; once again the EPA recommended against the project proceeding but was overruled and formal development approval granted by the Environment Minister. At the time of writing the project is under construction.

South West Yarragadee water supply development

The SWY sustainability assessment commenced soon after the completion of the Gorgon assessment, and although it was also a regulatory assessment process the approach taken was markedly different. The proponent was the Water Corporation of Western Australia, the government-owned water utility, and the proposal involved the extraction of 45 GL/day of groundwater from an aquifer approximately 300km south of Perth. From previous planning exercises, the Water Corporation had identified this source as the next most suitable water source to supply the city of Perth and connected areas. The purpose of the sustainability assessment was therefore not to compare this option with any others but to determine the most sustainable way to develop the resource. The starting point for the assessment was a conceptual design or a 'rubbery proposal' (Pope and Grace, 2006).

The Water Corporation worked closely with the Western Australian Government to develop the process steps and the governance structure for the sustainability assessment, drawing heavily from the lessons learnt from Gorgon. Key improvements introduced included: the assessment process commenced before the proposal was finalised, and therefore had elements of both internal and external sustainability assessment; clear sustainability objectives were established early in the process; a non-statutory Sustainability Panel (modelled on the Canadian Panel approach to environmental assessment under the Canadian *Environmental Assessment Act* 1992 – see Chapter 11) was formed to provide integrated advice to government alongside the environmental advice provided by the EPA through the statutory EIA process; and the social, economic and environmental implications of the proposal were reviewed within an integrated sustainability context by both the Sustainability Panel and the project team.

The assessment followed an iterative process, whereby the proposal was evaluated against the defined sustainability objectives, and modified if required to ensure a better performance against all the objectives (Strategen, 2006). Tensions between apparently competing environmental and social objectives were resolved by reframing the concept design. The result was a proposal that was demonstrably more sustainable than the original, and one which initially at least appeared to have more community support. Both the Sustainability Panel and the EPA recommended that the project proceed, albeit with extensive conditions attached to guide and monitor the development in order to ensure acceptable outcomes would be delivered (EPA, 2006a; Sustainability Panel, 2007).

However, although the sustainability assessment process up until this point was generally considered to have been successful, community opposition increased in the period leading up to the government decision. In response, the Premier of Western Australia rejected the proposal and in the same moment announced the construction and location of the Southern Seawater Desalination Plant (Perth's second desalination plant) in order to meet water supply demands (Carpenter, 2007). This was deeply ironic given that this decision was derived from an almost completely opaque process, in stark contrast with the rejected SWY proposal. That desalination plant was subjected to normal EIA and is nearing completion at the time of writing.

Internal proponent-led sustainability assessment

By definition, sustainability assessments undertaken by proponents to inform their own decision-making processes are *internal* sustainability assessments (Pope, 2006). Several different forms can be distinguished, one particularly common application being the evaluation of options (commonly site options) using techniques such as multi-criteria analysis (MCA) using criteria reflecting a range of sustainability considerations. Another distinct and interesting trend is that proponents are increasingly undertaking social impact assessments (SIAs) and developing social impact management plans (SIMPs) even though these are not required under Western Australian law. These are often made public, and sometimes incorporated into the proponent's statutory EIA documentation. It is worth highlighting here that the proponents themselves recognise the value of presenting their projects within a sustainability context.

Site selection sustainability assessment

In 2005 BHP Billiton voluntarily engaged in what can be classified as a sustainability assessment for the purposes of identifying a site to locate a new LNG plant for processing natural gas from the Scarborough field 250km offshore from the north-west coast of Western Australia (URS, undated).

The process commenced with the identification of all potentially suitable coastal locations within 400km of the gas field based on broad regional constraints. One of these sites was on Barrow Island, in the vicinity of the Gorgon gas development site, but this was rejected because it was considered to be contrary to the BHP Billiton Corporate Charter. The shortlist of eight potential sites was subjected to evaluation against a specified set of environmental, socio-economic and safety hazard risk factors, or criteria (but not including financial considerations). Transparency and community consultation and participation throughout the process were fundamental parts of the methodological approach (URS, undated).

Only once a favoured site emerged from the sustainability assessment process did BHP Billiton carry out an engineering cost–benefit analysis of the potential sites; it turned out that the chosen site was the most cost-effective solution. Following this sustainability assessment process, BHP Billiton then proceeded into the formal EIA process for the LNG plant at the chosen site which was subsequently approved and the project is currently being implemented.

More recently, the Northern Development Taskforce (NDT) appointed by the Government of Western Australia undertook a sustainability-oriented MCA process in 2007–2008 to identify an appropriate site for a proposed multi-user precinct for the processing of natural gas from the Browse Basin in Western Australia's Kimberley Region (NDT, 2008). If the precinct is established, individual project proponents will have the opportunity to develop projects within the precinct for the purpose of processing natural gas from the offshore Browse Basin. The precinct site selection process incorporated several stages, and the EPA provided strategic advice to government on the environmental implications of the final shortlist of four sites (EPA, 2008).

Similar MCA-based sustainability assessment processes have been undertaken for the purpose of selecting sites for public infrastructure such as water treatment plants and power transmission lines These processes have usually involved extensive community engagement, with community members often given the opportunity to weight the sustainability criteria in terms of their significance to the decision, as input into the MCA.

Other options evaluation sustainability assessment

Options other than site selection options have also been subject to sustainability assessment processes grounded in MCA. For example, the Water Corporation's *Water Forever* project utilised a two-tier sustainability assessment process to evaluate the sustainability of a range of water supply options for Perth as input to long-term planning (Water Corporation, 2008); and the

Sir James Mitchell Park tree planting project in which alternative landscape plans were subject to sustainability assessment to determine the most appropriate way for a municipal council to plant additional trees on a section of iconic parkland adjacent to the Swan River in accordance with a 2001 foreshore management plan (City of South Perth, undated).

The respective proponents of these projects had different incentives for integrating a form of sustainability assessment with their planning and decision-making processes. In the case of the Sir James Mitchell Park project, the City of South Perth's main goal was to demonstrate a robust and transparent decision-making process with opportunities for community involvement that would provide decision makers (in this case elected local councillors) confidence that the proposal to plant trees would deliver sustainable outcomes for the city, that the recommended landscape plan represented the most sustainable option, and that interested members of the community had provided input and the majority were supportive of the proposal (Pope and Klass, 2010).

In the case of the *Water Forever* process, however, the sustainability assessment contributed directly to the development of a portfolio of water supply options intended to provide Perth and surrounds with climate-change resilience over the next 50 years (Water Corporation, 2009). *Water Forever* involved extensive community consultation and the identification of a broad range of potential future water supply options. The 35 options included new water-source developments (e.g. surface and groundwater), desalination, wastewater recycling and reuse schemes, changing water use behaviour and individual water supply and reuse schemes (e.g. household rainwater tanks and backyard grey-water systems).

A two-step multi-criteria analysis (MCA) process, with environmental, social and economic criteria, was used to rank the water options in terms of their sustainability performance. The first step was a high level screening aimed at eliminating any options considered unsustainable with the second providing a ranking of the remaining options according to their overall sustainability performance, highlighting particular strengths and weaknesses. In this case the purpose was not to identify a single 'most sustainable' option but to understand the sustainability implications of a range of options from which a robust portfolio of options could be identified (Water Corporation, 2008).

Environmental and social impact assessments

Another trend that can be distinguished is that project proponents, particularly proponents of major resource projects, are increasingly voluntarily conducting SIAs as a supplement to their EIAs, despite the lack of statutory requirement to consider anything other than biophysical environmental impacts (and some related social impacts such as impacts caused by noise and dust). Technically this is a re-emergence given that a Social Impact Unit existed within the state government in the early 1990s (see Beckwith, 1994).

Two examples by proponents discussed in this chapter are ChevronTexaco's Wheatstone LNG Project (ChevronTexaco, 2010), and the Water Corporation of Western Australia's Southern Seawater Desalination Plant (GHD, 2008). Given that it is also common practice for proponents to include in their EIA documentation some analysis of the potential economic benefits of their project, proponents are effectively making a sustainability statement in their project documentation, which is released for public comment as part of the EIA process. The irony is that the EPA as the regulator to which this documentation is submitted then only assesses the biophysical component of the proposal to determine its environmental acceptability, leaving the social (and economic) dimensions with nowhere to go from a regulatory assessment and approvals perspective. In their initial assessment of site selection for the Browse LNG Precinct the EPA (2008, p.14) stated: 'there is no formal process in Western Australia for the assessment of socio-economic impact or indeed for their integration with environmental issues into a sustainability assessment'. For this particular assessment they highlighted some of the important social issues brought to their attention by participants in the assessment process 'to ensure that their importance is not lost' (EPA, 2008, p.14) but stated that it is only 'the environmental aspects of these analyses on which it is qualified to comment' (EPA, 2008, p.14).

The strategic assessment of the proposed Browse LNG Precinct, which is well advanced at the time of writing, is a good example of a proponent recognising that their 'social licence to operate' depends upon a robust consideration of social impacts and a public demonstration of how they intend to manage these impacts and deliver benefits to the local community. As discussed previously, the identification of a suitable site for the precinct was the responsibility of the government-appointed NDT, which utilised a form of MCA-based sustainability assessment. The proposed precinct on the site determined by the NDT process (James Price Point, approximately 60km north of the major tourist destination of Broome) is now subject to strategic assessment. The proponent in this case is the Western Australian Department of State Development (DSD) on behalf of the Government of Western Australia (for documentation on this project see http://www.dsd.wa.gov.au/8249.aspx). As might be expected, the potential social impacts of the proposed precinct are significant, particularly on the town of Broome and the Aboriginal communities close to the proposed site, especially since it would represent the first industrial development of this scale in the West Kimberley. Following what is arguably becoming standard practice, the proponent has undertaken a comprehensive SIA as part of its strategic assessment report. At the time of writing, DSD's strategic assessment report is being assessed by the Western Australian EPA (under the *EPAct* 1986) and the Federal Department of Environment, Water, Sustainability, Population and Communities (under the *Environment Protection and Biodiversity Conservation Act* 1999). The extent to which social impacts will be considered in their assessment and recommendations to the respective governments remains to be seen.

Procedural effectiveness

The absence of any formal requirement to conduct sustainability assessment in Western Australia means that each sustainability assessment process is developed on a case-by-case basis in response to an identified need or opportunity. Given that some of the case studies discussed in this chapter are examples of external, regulatory sustainability assessment (conducted for the purpose of determining whether or not a proposal should be approved) while others reflect internal processes informing project planning and development (conducted to select between options), it is not surprising that approaches to sustainability assessment have varied considerably. However, it can also be observed that each process incorporated some common steps: the identification of relevant sustainability issues (which in some cases were developed into clear objectives or criteria); assessment of the performance of the proposal or the options with respect to these issues; some level of community and stakeholder engagement; and a final decision by a nominated decision maker, either a regulator or internal to the organisation as appropriate.

The processes developed for the two case study examples of external, regulatory sustainability assessment (Gorgon and SWY) were structured around the statutory EIA process, which provided a robust and well-established structure upon which to build. However, the Gorgon case clearly demonstrated the limitations of simply adding a parallel strategic, economic and social 'stream' to a reactive EIA process. As discussed above, this approach left no opportunity to consider the dimensions of sustainability in an integrated fashion, and no scope to consider alternatives, particularly alternative locations for the development. While demonstrating 'comprehensiveness' in its scope, it performed poorly in terms of 'integratedness' and 'strategicness', two of the dimensions of best practice sustainability assessment identified by Hacking and Guthrie (2008).

In contrast, the SWY sustainability assessment process commenced sufficiently early in the project planning process to allow the findings of the assessment process to inform the project definition. Sustainability considerations were discussed by both the project team and the Sustainability Panel in a holistic and integrated fashion, and this led directly to an amended project concept that provided an opportunity to deliver on two sustainability objectives that had previously been in conflict. The Sustainability Panel also represented an innovative improvement over the Gorgon process with respect to governance, enabling the provision of a clear recommendation to Cabinet based upon a holistic, sustainability oriented assessment of the proposal. We conclude that the SWY process not only aligned with established EIA practice but demonstrated significant progress (from Gorgon) towards the evolution of an effective sustainability assessment process.

The two forms of proponent-driven sustainability assessment practice that have emerged, namely the use of MCA techniques with a sustainability orientation to distinguish between options during planning, and the undertaking of SIA and development of SIMPs, are based upon well-established methodologies.

Substantive effectiveness

The substantive effectiveness of sustainability assessment can be demonstrated by changes in process, actions or outcomes. With respect to process developments, we believe that the lack of stipulated process for sustainability assessment may be a strength, since it allows for flexibility in process design that demonstrates learning from experience. We believe that the SWY process was significantly better than the Gorgon process, for example, for reasons highlighted in the previous section, and know from our personal experience that the developers of the SWY process drew significantly upon the lessons learnt from Gorgon. It is unfortunate that political support for this experimentation, and for sustainability assessment in general, has waned, leaving us with no further examples of external regulatory sustainability assessment to explore.

It also appears likely that the lessons learnt from Gorgon, particularly the criticism directed at the process for failing to adequately address alternative sites, may have encouraged proponents to adopt sustainability assessment methodologies to support site selection processes. The Pilbara LNG site selection case study was one example, and the practice has become common in a range of organisations, including the state-owned power and water utilities.

There is also anecdotal evidence that the Gorgon case study prompted policy and process improvements beyond sustainability assessment practice. For example, ChevronTexaco was required as part of the sustainability assessment process to demonstrate 'net conservation benefit', effectively a biodiversity offset. At the time there was no policy in place outlining how this should be achieved, but within two years the EPA had released its draft *Position Paper on Environmental Offsets* (EPA, 2006b). Environmental offsets have now become a normal part of conventional EIA practice in Western Australia (e.g. Hayes and Morrison-Saunders, 2007) being incorporated into the *Environmental Impact Assessment Administrative Procedures* 2010 (*Government Gazette*, 26 November 2010, No. 223: 5979–6000, Perth, s2) as a way to counterbalance adverse residual environmental impacts remaining after other mitigation measures have been exhausted.

Similarly it can be argued that the rejection of the SWY proposal by government precipitated the highly transparent and inclusive planning initiative of *Water Forever*, which belatedly provided the strategic context for the SWY proposal.

Considering outcomes from a slightly different perspective, in the case of the Sir James Mitchell Park tree planting project the sustainability assessment provided structure and legitimacy to a decision-making process that would otherwise have been conducted behind closed doors. Based upon previous attempts to plant additional trees on the park, any proposal by the City of South Perth would have been strongly opposed by certain groups within the community and would have been unlikely to have been successfully implemented. Thus the sustainability assessment enabled quite a different outcome from what might otherwise have been expected.

Transactive effectiveness

The fact that sustainability assessment has been entirely voluntary in Western Australia, and yet practice is established, is testimony to its perceived value by proponents and regulators. Although we have limited ourselves to only a few examples in this chapter, we can point to other practice including some consultancy firms who have taken the lead and encourage or even 'require' their clients to adopt a sustainability assessment approach to what would otherwise be conventional EIAs (e.g. Elliott, 2008).

Attendance at the two Sustainability Assessment Symposia in Western Australia over the past few years (over 90 participants in each case) again demonstrates that sustainability assessment is alive and well in Western Australia, and that practitioners (proponents and consultants to proponents) are convinced of its value.

Involvement in sustainability assessment processes has in some cases also been personally rewarding for those involved, as discussed in Chapter 7 in relation to the SWY assessment.

Normative effectiveness

In terms of reversing prevailing unsustainable trends, it is perhaps unrealistic to expect too much from individual projects given some of the systemic pressures on the environment, society and economy but in a project-only context sustainability assessments including Gorgon, SWY and the proposed Browse LNG Precinct have attempted to deliver positive outcomes rather than just minimise adverse effects (notwithstanding that for Gorgon this was contentious because of the unique nature of Barrow Island such that no amount of mitigation or compensation was considered acceptable by the EPA (2003) in the event of a loss of biodiversity values on the island).

The site selection processes for the Pilbara LNG plant, Browse LNG Precinct and various public infrastructure projects have likely chosen the best possible option from a sustainability assessment perspective although there must inevitably be negative impacts associated with the actual development activity. Similarly *Water Forever* has set up a mechanism for enabling future water sources to be selected from a portfolio of choices rather than the narrow single option focus of SWY.

Integration is at the heart of sustainability assessment practice, and arguably its motivation. Government and proponents alike can see that it makes sense to consider proposals (or options) in a holistic way. Even in the Gorgon example, which was not successful in this respect, integration was an underpinning goal of the process.

All processes exhibited a reasonable degree of openness and transparency, although this is not to say that improvements in stakeholder engagement and empowerment cannot continue to be advanced. Where Gorgon was based in large part on consultation and opportunities for the public to comment on project documentation, many of the other case studies demonstrate considerable investment in active community engagement and participation in the decision and implementation of the sustainability assessment processes. This is to be expected, since one of the main incentives for assessing and communicating the sustainability implications of a proposal or option is to obtain social legitimacy.

Overall with respect to impact assessment practice in Western Australia it is too early to be able to lay claim to having 'turned the ship around' in terms of charting a course towards sustainability, but at least the emerging sustainability assessments have resulted in practitioners starting to ask the right questions that will help movement in this direction.

Pluralism

There is a long tradition of stakeholder consultation surrounding development projects in Western Australia associated with EIA practice. However the extent to which this rises above 'consultation' per se to more engaging forms of interaction and participation is highly variable.

Of the case studies discussed, Gorgon probably most represented a traditional Western Australian EIA approach where most of the public input was attained through formal public review stages (i.e. public submissions on development proposals that the proponent is expected to respond to). In other cases there was much greater evidence that proponents openly and actively consulted affected persons: for example many of the MCA processes offered community members an opportunity to weight the sustainability criteria.

Knowledge and learning

There is clear evidence that instrumental learning within individual sustainability assessments resulted in changes to project design and assessment practices. Conceptual learning was particularly evident in the SWY case where the development concept itself was revised as a result of the sustainability assessment, and we suggest that an integrated, holistic and collaborative approach to sustainability assessment that embraces the pluralism inherent in the process is essential to initiate this form of learning.

Without regulatory requirements to guide sustainability assessment practice there has been an opportunity to experiment with different designs and approaches, and Western Australian practitioners have reflected on what has been learned from attempting and carrying out sustainability assessment and taken the trouble to analyse it and to share experiences and theoretical ideas in published materials. Overall we would argue that a spirit of knowledge and learning prevails in Western Australian practice with sustainability assessment, notwithstanding that most of that learning probably resides principally with the relatively small number of practitioners (e.g. consultants and some senior officials within regulatory agencies) involved in the case studies and experiences to date. Not until sustainability assessment is widespread or 'mainstream' practice would we expect broader learning across the entire community.

Concluding remarks

The absence of regulatory requirements to undertake sustainability assessment has not hindered enlightened proponents and regulators from initiating their own processes and practices. The *ad hoc* nature of practice is, overall, best characterised as 'learning by doing'. We take a cautiously optimistic view here that practice and interest is growing over time. Early attempts at sustainability assessment initiated by the Government of Western Australia appear to have whetted the appetite for more recent proponent and consultant led approaches, which seek to realise the benefits of taking a more holistic approach to development planning and impact assessment activities.

The most recent emphasis has been on strategic assessment, which both the government and private sector appear to be keen to pursue. While from a regulatory perspective strategic assessment in Western Australia predominantly has an environmental emphasis (i.e. formally it occurs under the *EPAct*), there are signs that the scope is expanding into socio-economic issues too. What remains to be seen is whether sufficiently sophisticated governance mechanisms will be put in place to ensure that the socio-economic aspects of development in the State of Western Australia will be managed into the future.

Of course there remains much that can still be improved. While our 'scorecard' against the sustainability assessment framework criteria (summarised in Table 10.1) is generally positive, perhaps the ultimate test for future progress surrounds the six sustainability imperatives identified by Gibson in Chapter 1. For example, there is no real evidence yet that the sustainability assessments carried out in Western Australia can demonstrate that existing unsustainable trends are being reversed. However if nothing else the initial attempts at sustainability assessment in Western Australia have started to put important

Table 10.1 Sustainability assessment scorecard for Western Australia

Framework criterion	Questions asked	Western Australian perspective
Procedural effectiveness	*Have appropriate processes been followed that reflect institutional and professional standards and procedures?*	Sustainability assessment processes in Western Australia have been developed on a case-by-case basis reflecting context and evolving expertise, building upon well-established practices such as EIA, SIA and MCA.
Substantive effectiveness	*In what ways, and to what extent, does sustainability assessment lead to changes in process, actions, or outcomes?*	There is evidence that development proposals have been improved through the application of sustainability assessment processes during project planning. Furthermore, the practice of sustainability assessment itself has continued to evolve and mature as proponents and regulators have learnt from others' experiences. Sustainability assessments have also identified policy and strategic gaps that have subsequently been addressed.
Transactive effectiveness	*To what extent, and by whom, is the outcome of conducting sustainability assessment considered to be worth the time and cost involved?*	The fact that proponents and regulators are volunteering to engage in sustainability assessments in the absence of any legal requirement to do so strongly implies that they see benefit from taking such an approach and therefore the cost and time investment is worthwhile.
Normative imperatives	*In what ways, and to what extent, does the sustainability assessment satisfy the listed normative imperatives?*	It is still too early to judge just how effective sustainability assessment practice in Western Australia is with respect to the six normative principles. While having open and engaging processes is a normal part of any impact assessment activity, and there is increasing respect for the complexity and context within which sustainability assessment takes place, challenges remain with integration, dealing with trade-offs and demonstrating that mutually reinforcing gains will be delivered by development activity that will reverse prevailing unsustainable trends.
Pluralism	*How, and to what extent, are affected and concerned parties accommodated into and satisfied by the sustainability assessment process?*	Communities are increasingly demanding that they are involved and have influence in sustainability assessment, and the notion of 'social licence to operate' is understood and respected by proponents and regulators. However community engagement practices can still be significantly enhanced to ensure movement from 'consult and comment' approaches to active engagement and empowerment.
Knowledge and learning	*How, and to what extent, does the sustainability assessment process facilitate instrumental and conceptual learning?*	There is clear evidence of instrumental learning, and in some cases conceptual learning, where sustainability assessment has directly influenced the development of proposals. An integrated, holistic and collaborative approach to sustainability is essential for the conceptual learning potential of sustainability assessment to be realised. Conceptual learning about sustainability assessment itself is also continually evident within the practitioner community.

sustainability issues into the mind-set of practitioners and the public and attempts are being made to better adopt an integrated approach, address trade-offs and engage more effectively with the community and affected parties. This is an important start and Western Australia has a valuable foundation upon which to continue building effective sustainability assessment processes and practice.

References

Allen Consulting Group (2003) *Proposed Access to Barrow Island for Gas Development: Advice on Social, Economic and Strategic Considerations. A Report to the WA Department of Industry and Resources.* Perth, Allen Consulting Group. [Online] Available at: <http://www.allenconsult.com.au/publications/view.php?id=283> (accessed 13 September 2011).

Bache S, Bailey J, Evans N (1996) 'Interpreting the Environmental Protection Act 1986 (WA): social impacts and the environment refined', *Environmental Planning and Law Journal*, 13, 487–492.

Beckwith J (1994) 'Social impact assessment in Western Australia at a crossroads', *Impact Assessment*, 12(2), 199–213.

Carpenter A (2007) *Second Seawater Desalination Plant to be State's Next Major Water Source.* Ministerial Media Statement 15/7/07. [Online] Available at: <http://www.mediastatements.wa.gov.au/Pages/Results.aspx?ItemID=128275> (accessed 13 September 2011).

ChevronTexaco (2010) *Draft Environmental Impact Statement/Environmental Review and Management Programme for the Proposed Wheatstone Project: Executive Summary.* ChevronTexaco, Perth, Western Australia. [Online] Available at: <http://www.chevronaustralia.com/Libraries/Chevron_Documents/Wheatstone_Draft_EIS_ERMP_Executive_Summary.pdf.sflb.ashx> (accessed 13 September 2011).

City of South Perth (undated) *Sir James Mitchell Park Tree Planting Project Draft Sustainability Assessment Report* (City of South Perth, Western Australia) [Online] Available at: <http://old.southperth.wa.gov.au/sustainability/pdf/DraftSustAssessReport_JP_v6Final.pdf> (accessed 19 July 2011).

Conservation Commission of Western Australia (2003) *Biodiversity Conservation Values on Barrow Island Nature Reserve and the Gorgon Gas Development: Advice to Government from the Conservation Commission of Western Australia.* (Crawley, WA: Conservation Commission of Western Australia).

DFAT – Department of Foreign Affairs and Trade (2010) *Western Australia Fact Sheet,* Government of Australia. [Online] Available at: <http://www.dfat.gov.au/geo/fs/wa.pdf> (accessed 7 September 2011).

DoIR – Department of Industry and Resources (2004) *Draft Retrospective Review on the In Principle Agreement for Considering Access to Barrow Island for the Processing of Gorgon Gas: Consolidated Summary Report* (unpublished) (Perth, Western Australia: DoIR).

DSD – Department of State Development (2010) *Browse Liquified Natural Gas Precinct Strategic Assessment Report* (draft for public comment) December 2010 Part 1 Executive Summary, DSD, Western Australia. [Online] Available at: <http://www.dsd.wa.gov.au/8249.aspx> (accessed 19 July 2011).

Elliott P (2008) *Sustainability Assessment Challenges and Lessons* (presented at Sustainability Assessment Symposium 2008). [Online] Available at: <http://integral-sustainability.net/wp-content/uploads/sas2-1-elliott.pdf> (accessed 21 July 2011).

EPA – Environmental Protection Authority (2003) *Environmental Advice on the Principle of Locating a Gas Processing Complex on Barrow Island Nature Reserve (Gorgon Venture)* (EPA Bulletin No. 1101) (Perth: EPA).

EPA – Environmental Protection Authority (2006a) *South West Yarragadee Water Supply Development (Water Corporation), Report and Recommendations of the Environmental Protection Authority* (EPA Bulletin No. 1245) (Perth: EPA). [Online] Available at: <http://www.epa.wa.gov.au/EPADocLib/2412_Bull1245%20Yarra%20ermp%2081206.pdf> (accessed 13 September 2011).

EPA – Environmental Protection Authority (2006b) *Position Paper No. 9: Environmental Offsets* (Perth: EPA). [Online] Available at: <http://www.epa.wa.gov.au/docs/1863_PS9.pdf> (accessed 13 September 2011).

EPA – Environmental Protection Authority (2008) *Kimberley LNG Precinct Review of Potential Sites for a Proposed Multi-user Liquefied Natural Gas Processing Precinct in the Kimberly Region, Advice of the EPA to the Minister for Planning (as the Minister for Environment's delegate) under Section 16(e) of the EPAct 1986*, Report 1306, (Perth: EPA).

Government of Western Australia (2002) *Focus on the Future: the Western Australian State Sustainability Strategy Consultation Draft* (Perth: Department of the Premier and Cabinet).

Government of Western Australia (2003) *Hope for the Future: The Western Australian State Sustainability Strategy* (Perth: Department of the Premier and Cabinet). [Online] Available at: <http://www.dec.wa.gov.au/content/view/3523/2066/> (accessed 21 July 2011).

GHD (2008) *Water Corporation: Report for the Proposed Southern Seawater Desalination Project – Social Impact Assessment.* [Online] Available at: <http://www.watercorporation.com.au/_files/PublicationsRegister/15/SSDP_GHD_06_05_08_MASTER_COPY.pdf> (accessed 13 September 2011).

Hacking T and Guthrie P (2008) 'A framework for clarifying the meaning of triple bottom-line, integrated, and sustainability assessment', *Environmental Impact Assessment Review*, 28(2–3), 73–89.

Hayes N and Morrison-Saunders A (2007) 'Effectiveness of environmental offsets in environmental impact assessment: practitioner perspectives from Western Australia', *Impact Assessment and Project Appraisal*, 25(3), 209–218.

Independent Review Committee (2002) *Review of the Project Development Approvals System: Final Report* (Perth: Government of Western Australia).

NDT – Northern Development Taskforce (2008) *Final Site Evaluation Report* (Perth: Government of Western Australia). [Online] Available at: <http://www.dsd.wa.gov.au/documents/000269V04.GARY.SIMMONS.pdf> (accessed 13 September 2011).

Newman P (2006) 'Sustainability assessment'. In Marinova D, Annandale D and Phillimore J (eds) *The International Handbook on Environmental Technology Management.* (Cheltenham: Edward Elgar Publishing).

Pope J (2006) 'Editorial: what's so special about sustainability assessment?', *Journal of Environmental Assessment Policy and Management*, 8(3), v–ix.

Pope J (2007) *Facing the Gorgon: SA and Policy Learning in Western Australia*, PhD thesis, Murdoch University. [Online] Available at: <http://researchrepository.murdoch.edu.au/264/> (accessed 7 July 2011).

Pope J and Grace W (2006) 'Sustainability assessment in context: issues of process, policy and governance', *Journal of Environmental Assessment Policy and Management*, 8(3), 373–398.

Pope J and Klass D (2010) 'Decision quality for sustainability assessment' *IAIA10 Conference Proceedings, The Role of Impact Assessment in Transitioning to the Green Economy 30th Annual Meeting of the International Association for Impact Assessment 6–11 April 2010, International Conference Centre Geneva – Switzerland*. [Online] Available at: <http://www.iaia.org/iaia10/proceedings/submitted-papers.aspx> (accessed 19 July 2011).

Pope J, Morrison-Saunders A, Annandale D (2005) 'Applying sustainability assessment models', *Impact Assessment and Project Appraisal*, 23(4), 293–302.

Strategen (2006) *South West Yarragadee Water Supply Development: Sustainability Evaluation/Environmental Review and Management Programme (ERMP)*. Report prepared for Water Corporation, Perth (Subiaco: Strategen).

Sustainability Panel (2007) *Sustainability Assessment of the South West Yarragadee Water Supply Development*. Report prepared for the State Water Council, Department of the Premier and Cabinet, Perth, Western Australia. [Online] Available at: <http://www.water.wa.gov.au/PublicationStore/first/73423.pdf> (accessed 13 September 2011).

URS Australia Pty Ltd (undated) *Pilbara LNG Project Site Selection Study* (Perth: BHP Billiton).

Water Authority of Western Australia (1995) *Perth's Water Future: A Water Supply Strategy for Perth and Mandurah to 2021 (With a Focus to 2010)* (Leedervilles: Water Authority). [Online] Available at: <http://www.watercorporation.com.au/_files/PublicationsRegister/12/perth_water_future.pdf> (accessed 13 September 2011).

Water Corporation (2008) *Water Forever Sustainability Assessment*, December 2008 (Leederville: Water Corporation of Western Australia). [Online] Available at: <http://www.thinking50.com.au/go/publications> (accessed 19 July 2011).

Water Corporation (2009) *Water Forever: Towards Climate Resilience*, December 2008 (Leederville: Water Corporation of Western Australia). [Online] Available at: <http://www.thinking50.com.au/go/publications> (accessed 19 July 2011).

Wood C (1994) 'Lessons from comparative practice', *Built Environment*, 20(4), 332–344.

11 Sustainability assessment in Canada

Robert B. Gibson, University of Waterloo

> In preparing for public hearings, the Proponent, Interveners and other participants should be aware that the Panel will evaluate the specific and overall sustainability effects of the proposed project and whether the proposed project will bring lasting net gains and whether the trade-offs made to ensure these gains are acceptable in the circumstances.
>
> Mackenzie Gas Project Joint Review Panel (2005)

Introduction

Canada is a big and mostly fortunate country. It has plenty of space and considerable capacity for a wide range of experiments with many things, including sustainability assessments. For a variety of reasons, Canada does not have a formal sustainability-based assessment regime, at least not one that can easily be tested against the criteria set out in this book. But it does have a long and illuminating record of *de facto* sustainability assessments, only some of which were initiated under environmental assessment law. The discussion to follow will survey the range and high points of Canadian experience with sustainability assessments, identify strengths and limitations, and consider what broader lessons may be drawn from experience that is, inevitably, somewhat peculiar to the country involved.

Sustainability assessment as public practice

Humans have been immigrating to Canada for 20,000 years or more, bringing and building a diversity of cultures in a wide variety of climates and ecologies. The earliest arrivals, peoples now considered Aboriginal, were and mostly remain more inclined to assess options from a sustainability perspective than the European and other traditions that arrived later. The usual modern ideas that are centred on economic growth through the conquest of nature have prevailed in Canada as almost everywhere else. But there have always also been detectable counter positions – appreciations of place, community and beauty that maintained longer and richer perspectives. And even in such a rich and generally advantaged country, the costs of unsustainable practices and needs for more farsighted and broadly beneficial undertakings have been visible to anyone who cares to notice.

It should not be surprising, then, to find examples of sustainability-based deliberation in many forms and venues. When lobster fishers operating out of a small harbour on Cape Breton Island, Nova Scotia, consider how best to maintain their customary means of allocating access to preserve the resource and community incomes fairly over the long term, they are engaged in a form of sustainability assessment. So are farmers and conservationists in Norfolk County, Ontario, discussing how to support agricultural livelihoods as well as ecological rehabilitation through payments for ecological goods and services. The multi-stakeholder development of Forest Stewardship Council certification rules for particular Canadian forests involves sustainability assessment. An undergraduate student group that is organizing a cooperative lunch counter serving local and fair trade products is carrying out a series of sustainability-based assessments.

I could go on. Probably I could fill this chapter with Canadian examples and merely scratch the surface. A roughly similar case might be made for most parts of the world. Probably most exemplary initiatives that seek lasting gains across a range of interdependent objectives arose from some form of sustainability assessment and may be applying that approach still through iterative learning. The words 'sustainability' and 'assessment' may not have been used. The participants may not have seen their effort as a special process. Perhaps no single decision or consequent undertaking was involved, and no particular legislated obligation or set of formal procedures ruled. Typically these assessments were and are phenomena of governance rather than government, of voluntary collaborations rather than formal authority. They are not usual subjects for evaluations of assessment processes and no one to my knowledge has attempted even to define the main categories, much less assemble a reasonably comprehensive list of the most notable cases. But while it is not possible in this space to present a reasonably credible and comprehensive review of these initiatives, it is crucial to recognize their significance – their evident number and diversity, their presence at every scale from the neighbourhood to the nation, their substantial (if more or less seriously imperfect) adherence to the basics of sustainability assessment, and most importantly their base in the recognized demands of actual circumstances rather than the imperatives of law and policy.

For the purposes of this chapter on sustainability assessment in Canada, I will focus on the experience to date in formal processes driven by government authority of some sort. As will be discussed below, the formal versions of sustainability assessment in Canada have also been numerous and diverse. They too have been mostly *ad hoc* and in many cases have happened despite, rather than because of, what was established by the relevant authorities.

In this they have reflected the character, and indicated the influence of the less formal initiatives. It is impossible to determine just how much the evolution of formal sustainability assessments in Canada owes to the proliferation of less formal, collaborative and circumstance-driven sustainability-based deliberations. But certainly a chapter on sustainability assessment in Canada properly begins with a salute to the innovators outside the usual assessment circle.

Evaluating formal sustainability assessments in Canada

In the introductory chapter of this book, I argued that the core of what sustainability assessment should deliver lies in six imperatives: sustainability assessment must

- aim to reverse the prevailing (unsustainable) trends
- integrate attention to all of the key intertwined factors that affect sustainability
- seek mutually reinforcing gains
- minimize trade-offs
- respect the context
- be open and broadly engaging.

These imperatives are complemented by other key considerations in the editors' Chapter 8, which sets out a more complete set of criteria emphasizing as well matters of procedural completeness, efficiency, learning, accommodation of interests, and actual delivery of substantive improvements.

Taken together, these criteria set a high standard. Whether any of the informal and collaborative initiatives discussed above meet this standard is an open question. In the realm of more formal, government-led law and practice involving assessments of some sort, it is doubtful that any existing regime comes close to providing a clear set of requirements and procedures for consistently effective, sustainability-based assessment. There is room for debate here because no one has attempted a comprehensive analysis of the many regimes that should be considered.

In Canada, responsibilities for sustainability-related issues are divided and shared among federal, provincial, territorial, Aboriginal, and municipal authorities. Some of the resulting complexities are evident in the multiplicity of law-based environmental assessment processes. The *Canadian Environmental Assessment Act* (Government of Canada, 2011) is accompanied by assessment legislation in each of the ten provinces and three territories, plus several more based in Aboriginal land claim agreements. Additional strategic level assessment requirements are established within law-based regimes for urban and regional planning and for many particular sectors (e.g. management of forest lands, electrical energy systems, and telecommunications). And special project-level assessment requirements of various kinds are set out in laws concerning nuclear facilities, aggregates extraction, exports financing, and a host of other matters. No two of these regimes are the same. Not surprisingly, the challenges of coordination have led to many calls for harmonization and simplification. At the same time, however, some of the most salutary advances in public policy substance and process have come from initiatives of interjurisdictional collaboration, combining existing processes and/or establishing new joint mechanisms. That has been the story in sustainability assessment.

None of these many formal planning and/or assessment regimes in Canada include explicit recognition of the six core imperatives of sustainability assessment or are designed to meet more than a few of the other criteria set out in this book. The most that can be claimed is that over the past 40 years, a promising series of individual assessments or assessment-like initiatives have explored and demonstrated some of the rich possibilities of sustainability-based deliberation and evaluation. While no initiatives would satisfy all of the criteria, the cases collectively represent the gradual emergence of sustainability assessment practice in Canada and the best examples are, despite their imperfections, probably at the leading edge of practice in the world.

The following section provides brief accounts of seven initiatives that were, effectively, sustainability assessments. The seven were undertaken by several different authorities, sometimes in collaboration, and relied on a variety of legal foundations, including planning, resource management, and public inquiries law as well as environmental assessment legislation. All of these assessments were exceptional and most involved special arrangements to deal with a particularly challenging or controversial topic. The first three begin with the initial, powerful precedent of the Mackenzie Valley Pipeline Inquiry, and include a strategic-level forest-sector case under environmental assessment law, and a regional urban growth management case under planning law. They represent a larger diversity of examples with similar fundamentals. The last four cases provide the nearest Canadian approximation of an emerging line of practice. All are major project assessments (one is effectively strategic as well) where an explicit sustainability test was applied by independent review panels with public hearings, and all of them rest on provisions of the *Canadian Environmental Assessment Act* combined with the requirements of at least one other jurisdiction. These cases are characterized by the gradual evolution of sustainability assessment practice to emphasize rigorous comparative evaluation of alternatives in light of a well-developed and quite comprehensive set of sustainability criteria.

Seven sustainability assessments

The Mackenzie Valley Pipeline Inquiry (1974–1977)

In Canada in the 1970s, governments faced rising environmental awareness, demands for more transparent and participative decision making, and belatedly recognized Aboriginal rights. These influences came together in controversies surrounding a proposed multi-billion dollar pipeline to carry natural gas from Alaska and the Canadian western arctic, up the Mackenzie Valley in the Northwest Territories to markets in the south. In response, the federal government appointed Mr Justice Thomas Berger of the British Columbia Supreme Court to carry out a special public inquiry.

Berger's formal mandate was to examine the potential social-economic and cultural as well as biophysical effects of the proposed project and to recommend suitable terms and conditions for approving the project. However, he recognized that decisions on the pipeline would also be decisions about the future of a large portion of the Canadian north and that two competing visions were in play. For the project proponents, the north was a resource frontier for the industrial economy; for the largely Aboriginal residents, the north was a homeland. Berger's inquiry therefore centred not just on the effects of the pipeline and other developments it would induce, but also on whether and how the two visions could be reconciled (Gamble, 1978; Dacks, 1981; Page, 1986).

Assisted by great public interest across Canada, Berger used his independent authority to hold quasi-judicial public hearings with technical sessions for experts and community sessions in every settlement along or near the pipeline route. He also introduced intervenor funding, the provision of public funds to facilitate effective participation by stakeholders who have relevant perspectives and interests but who lack adequate resources (Gamble, 1978, pp.949–950).

Berger's final report, *Northern Frontier, Northern Homeland*, compared pipeline options, evaluated potential effects and uncertainties, and most notably, recommended a ten-year delay of project approval to allow for negotiation of land claims agreements between the governments and Aboriginal groups (Berger, 1977). Government authorities agreed, though it is not clear whether they were moved by Berger's arguments so much as by rising doubts about the project's potential economic viability in the face of competition from cheaper gas supplies near existing distribution systems in Alberta (Robinson, 1983). The pipeline proposal did not re-emerge for another quarter century.

Ontario's class environmental assessment of timber management on crown lands (1987–1994)

The forest industry in Canada has long been economically important, politically influential and, in many places, heavily reliant on harvesting from publicly owned 'crown lands' under provincial authority. Traditionally, the provinces have exercised their planning and permitting control in close collaboration with the industry, treating forests mostly as a source of timber and fibre, and favouring foreseeable economic priorities over resource sustainability.

In Ontario, the inevitably rising conflicts with other forest users came to a head in the 1980s. By an accident of timing, the venue for the policy debate was the provincial environmental assessment process. The Ontario *Environmental Assessment Act* applies automatically to all provincial undertakings, plans as well as projects, unless formally exempted from assessment. To avoid multitudes of individual assessments of particular plans for access roads, harvesting, renewal and maintenance, the province invented a strategic level 'class environmental assessment' mechanism. It would address overall management issues and set out processes for developing and approving more specific plans for each of the 114 forest management units in the province.

In 1987, after more than a decade of delays, the Ministry of Natural Resources submitted its *Class Environmental Assessment for Timber Management on Crown Lands in Ontario* for quasi-judicial public hearings before a panel of the Environmental Assessment Board. Critics found the Ministry's class assessment document vague, narrowly focused on timber priorities and unlikely to maintain forest values in perpetuity. But it opened an inquiry into broad alternatives for forest planning and management, considering the full range of social, economic and cultural as well as biophysical effects, the implications for all forest uses and users, and the lasting maintenance of the resource (Dunster *et al.*, 1989).

The hearings lasted nearly four years. Virtually everyone found them unacceptably long, difficult and costly. But they played a major role in inducing a substantial policy shift that may not have been accomplished otherwise. Throughout all the previous years, Ministry officials had held firmly to their traditional focus on supplying the forest products industry, rather than integrated multi-purpose forest management engaging all forest interests. By the end of the hearings, the Ministry had adopted policy reforms recognizing non-timber uses of forest lands and introducing a consultative approach to forest planning. In April 1994, when the Board finally issued an approval with a long list of detailed conditions (EAB, 1994), the province had a new 'Policy Framework for Sustainable Forests' and was about to pass a new *Crown Forest Sustainability Act*, responding to the issues raised at the hearings.

While the Ontario timber management case was not formally an exercise in sustainability assessment, sustainability questions underlay all of the deliberations. The approach taken was messy and aggravating. It nevertheless demonstrated the power of assessment processes to encourage sustainability-oriented reform of basic policies and processes in the face of stiff proponent-resistance.

Development of an urban growth management strategy for British Columbia's Capital Regional District (1996–2003)

British Columbia's Capital Regional District (CRD) includes the city of Victoria and 15 adjacent municipalities and electoral districts at the south end of Vancouver Island. By the early 1990s the region was facing significant growth controversies. The population had been expanding quickly but the region, almost encircled by water, had limited space for urban expansion and strong public support for the remaining green spaces, recreational areas and agricultural lands. Unfortunately, no mechanism for effective response was immediately available. British Columbia does not have a tradition of strong regional governance and the CRD as a regional authority is a creature of its independent-minded constituent municipalities.

In 1995 the British Columbia legislature passed a new planning law encouraging municipalities with increasing populations to prepare Regional Growth Strategies (BC, 1996). As means of coordinating municipal action on regional issues, the strategies would be powerful. The municipalities' Official Community Plans would have to comply with approved regional strategies, as would infrastructure financing and other agreements with the province. In addition, the strategies would facilitate pursuit of sustainability objectives. Mandatory strategy contents covered 14 goals, including reducing urban sprawl, protecting environmentally sensitive areas, providing affordable housing and decreasing pollution (BC, 1996: s.849(2)).

Development of the CRD growth strategy took seven years. The process followed conventional rational planning steps – information gathering, trend analysis, priorities identification, scenario comparisons, and final negotiation of the details of the preferred option. It was also consultative, encouraged public involvement at successive stages, and was underpinned by the province's sustainability-based growth strategy goals. While much of the initiative and direction came from municipal leaders and the regional planning staff, key roles were played by a public advisory committee and a diversity of residents and citizens' groups. A key early step was depiction and publication of the business-as-usual scenario: the overall built-out effect of continued growth following the municipal plans then in place. Strongly negative public reaction to this scenario set the stage for a more motivated examination of alternatives that would preserve desired qualities and promise a generally more desirable future (Boyle *et al.*, 2004).

Negotiating the details of the strategy, especially concerning matters related to the placement and firmness of the urban containment boundary and the particular locations for densification within the boundary, was particularly difficult. Years of discussion, mediation and compromise were needed before the CRD Regional Growth Strategy was finally approved and adopted as a regional by-law in 2003. The result has not ended growth tensions and is unlikely to deliver a model for urban regional sustainability. But the sustainability-based CRD Strategy process and result represent a significant transition to a substantially different approach to urban growth, with important implications not just for planning policies and practice but also for associated infrastructure options, building design, services delivery, financing priorities, and a host of other particulars.

The Voisey's Bay mine and mill environmental assessment (1997–2002)

Voisey's Bay on the north coast of Labrador is in the intersecting traditional territories of the Aboriginal Innu and Inuit. It is also subject to the overlapping authority of the Canadian federal government and the provincial government of Newfoundland and Labrador. In 1997, despite or perhaps because of a history of conflict (Gibson, 2006; O'Faircheallaigh, 2006), these four jurisdictions agreed to establish a joint panel to guide and review the environmental assessment of a nickel mine and mill, plus an associated port and marine shipping, proposed by a subsidiary of Inco Ltd (Government of Newfoundland and Labrador *et al.*, 1997). As with other such panel-level environmental assessments of major undertakings in Canada, the process for the Voisey's Bay Panel involved issuing guidelines for the proponent's preparation of an Environmental Impact Statement, receiving and reviewing the general adequacy of the submission, holding public hearings to consider the proposed project, carrying out a final review in light of the evidence received, and preparing a report with recommendations to the relevant governments.

The five-member Panel's terms of reference were broad, incorporating attention to a comprehensive set of human and biophysical factors, welcoming traditional ecological knowledge, and recognizing cumulative effects, beneficial effects, and lasting effects on renewable resources (Government of Newfoundland and Labrador *et al.*, 1997). 'Sustainability' was not mentioned. The Panel, however, interpreted its mandate as effectively requiring consideration of 'the extent to which the Undertaking may make a positive overall contribution towards the attainment of ecological and community sustainability, both at the local and regional levels' with attention to the preservation of ecosystem integrity, the rights of future generations, and 'the attainment of durable and equitable social and economic benefits' (Voisey's Bay Panel, 1997, s.3.3). In doing so, the Panel became the first in Canada to adopt and impose an explicit 'contribution to sustainability' test in the review of the proposed undertaking.

Mining is a counter-intuitive subject for sustainability expectations. Orebodies are depletable resources and mines are typically associated with immediate gains and permanent damage rather than lasting foundations for wellbeing. The main Voisey's Bay orebody, 'the Ovoid', was exceptionally rich and conveniently close to tidewater, but it was small. The 20,000 tonnes/day mill proposed by Inco would have been able to exhaust the Ovoid in about seven years. For the Panel, however, the key issue was whether and how the project could be undertaken so that it would leave a positive legacy. In particular the Panel was interested in how the project life could be extended to provide a longer stream of benefits and allow more time and opportunity to build capacities and options for viable livelihoods when the mine closed (Voisey's Bay Panel, 1999, s.2.3). After its initial review of the submitted environmental impact statement, the Panel required additional information on possible alternative rates of ore extraction. This concern rose again in the Panel's public hearings in ten Labrador communities and in the provincial capital, and were central in the Panel's recommendations.

In its final report, released in March 1999, the Panel concluded that the project should be authorized subject to terms and conditions that the Panel set out in 107 recommendations (Voisey's Bay Panel, 1999, s.18). The recommendations covered a wide range of social, economic and ecological matters, but focused chiefly on means of extending the lifetime of the project and ensuring a flow of opportunities and potentially lasting benefits to the Innu and Inuit communities of the region. The company initially resisted reducing the capacity of the mill to ensure a longer project life. In the end, however, Inco agreed to build a 6,000 tonnes/day mill, less than a third of the size of the one originally proposed, anticipating a project life of at least 30 years (Inco Limited, 2002).

The results probably fall short of ensuring durable livelihoods after the mining ends and the Panel's approach did not encompass many of the national- and global-scale sustainability issues surrounding mining (Green, 1998). The Panel's pioneering sustainability-based assessment did, however, lead to remarkable agreement among Aboriginal and government interests that had long histories of conflict (Gibson, 2006), and set a contribution to sustainability precedent to be followed by subsequent panels established in part under the *Canadian Environmental Assessment Act.*

Whites Point quarry and marine terminal environmental assessment (2004–2007)

Late in 2004, the province of Nova Scotia and the Canadian federal government agreed to appoint a three-member panel to review a proposal for a large basalt quarry and associated shipping facilities, which had stirred considerable local opposition. The site was at Whites Point on Digby Neck, a scenic peninsula on the Bay of Fundy.

The quarry proponents – Bilcon, a US company based in New Jersey – anticipated a 50-year project, with local employment benefits, associated income tax gains for governments, and progressive rehabilitation of the site. Critics feared adverse effects on tourism and fishing, additional stresses on endangered whales and other marine species due to the increased ship traffic, minimal economic benefits, loss of tranquility, and a permanently scarred landscape (Whites Point Panel, 2007, pp.27–85).

With a federal–provincial mandate similar to the one provided in the Voisey's Bay case, the Whites Point Panel issued assessment guidelines that incorporated a contribution to sustainability test, using language borrowed from the Voisey's Bay guidelines. After a lively round of local hearings, the Panel undertook an analysis focused on compliance with the Panel's guidelines, including their guiding principles. The Panel gave particular attention to project viability, community sustainability, and the nature and distribution of benefits and burdens (Whites Point Panel, 2007, pp.13–14, 86–100).

The Panel concluded that the project 'would not make a net contribution to sustainability', that the economic gains would accrue mostly to the proponent at the expense of long-term qualities and sustainable community economic development opportunities consistent with the core values of the community, and that the project should not be approved (Whites Point Panel, 2007, p.101). As well, the Panel addressed a set of strategic-level concerns arising from its inquiry, including the evident need for anticipatory coastal zone planning. The federal and provincial authorities agreed to reject the proposed project and to consider the broader recommendations (Government of Nova Scotia, 2007).

Kemess North copper-gold mine environmental assessment (2005–2007)

The Kemess North Joint Review Panel appointed by the federal government and the provincial government of British Columbia was the third formal assessment panel in Canada to receive and apply an explicitly sustainability-focused mandate. Like its Voisey's Bay predecessor, the Kemess North Panel reviewed a proposal for a mine with a short life expectancy (11 years of anticipated mine operation) and substantial Aboriginal interests at stake.

The Kemess North mine, in north central British Columbia, was proposed as an expansion of an existing mine (Kemess South), six kilometres away. The new mine would benefit from use of the existing mine's infrastructure and would extend mine employment (475 jobs) and other social and economic benefits. In addition, however, the project involved dumping several hundred million tonnes of acid-generating mine tailings and waste rock into a natural lake that is spiritually significant to local First Nations.

To weigh the pros and cons, the Panel adopted a sustainability assessment framework drawing from earlier documents prepared by the international mining sector and the provincial government (Kemess North Panel, 2007, pp.233–234). The framework applied five 'sustainability perspectives: environmental stewardship; economic benefits and costs; social and cultural benefits and costs; fairness in the distribution of benefits and costs: and present versus future generations' (Kemess North Panel, 2007, p.xi). In the final chapter of its report, the Panel considered the effects of the proposed project from each perspective (Kemess North Panel, 2007, pp.234–245).

The Panel concluded that 'the project in its present form would not be in the public interest' because the recognized economic and social benefits would be transient and 'outweighed by the risks of significant adverse environmental, social and culture effects, some of which may not emerge until many years after mining operations cease' (Kemess North Panel, 2007, p.245). Central among the long-term adverse effects concerns were loss of the valued natural lake and the legacy of tailings management obligations, perhaps lasting thousands of years, to prevent acidification and other damage to downstream waters.

The federal and provincial authorities accepted the Panel's recommendations and denied the proponent's application.

Mackenzie Gas Project environmental assessment (2004–2009)

In August 2004, 30 years after the Berger Inquiry began, federal, territorial and Aboriginal authorities jointly announced the appointment of a new, seven-member environmental assessment panel to review a resurrected Mackenzie gas gathering and pipeline project (CEAA, 2004; MVEIRB *et al.*, 2004). Building not only on Berger's work but also on the intervening decades of learning about regional-scale assessment, northern development, and applied sustainability, the Joint Review Panel for the Mackenzie Gas Project has provided Canada's most advanced example of assessment applying a contribution to sustainability test (Gibson, 2011).

The Mackenzie Panel addressed an exceptionally challenging version of project-based assessment. The project as filed by a hydrocarbon industry consortium was for a $16.2 billion package involving development of three gas fields in the Mackenzie Delta area, associated gas gathering facilities, and a 1200km pipeline up the Mackenzie Valley. The significant impacts, however, would also include those of additional, induced developments. While the initial three gas fields were expected to deliver 0.83 billion cubic feet per day (Bcf/d) of gas, the pipeline was designed to carry 1.2 Bcf/d immediately and to accommodate 1.8 Bcf/d through the addition of more heater and compressor stations. Some scenarios presented to the Panel anticipated even higher gas throughput and accordingly greater cumulative impacts, positive and negative, from more gas field and related infrastructure development, more revenues, more opportunities, and more stresses on ecological, social and administrative capacities (Mackenzie Panel, 2009, chap.3). Effectively, the case was a strategic assessment in the guise of a project assessment review.

Unlike the earlier panels, the Mackenzie Panel did not create its own guidelines for the preparation of the environmental impact statement. Instead the guides were provided in the Panel's terms of reference from the three governments. Also for the first time, the government-established mandate explicitly established 'contribution to sustainability' along with respect for traditional knowledge, land claims and treaties, diversity and the precautionary approach, as fundamental principles for the assessment (IGC *et al.*, 2004, p.4). Like the other panel cases, the subject was a non-renewable resource undertaking that could not itself be sustainable and could contribute to sustainability only through a positive legacy.

The Panel interpreted its mandate in a clear statement of its sustainability test (see the quote that begins this chapter) and established a detailed analytical framework based on 36 key issues in five core categories that were meant to cover the full suite of requirements for progress towards sustainability (Mackenzie Panel, 2009, esp. chaps 5 and 19):

- cumulative impacts on the biophysical environment
- cumulative impacts on the human environment
- equity impacts (fair distribution of benefits and risks)
- legacy and bridging impacts
- cumulative impacts management and preparedness (capacities for managing the risks and opportunities).

In a process involving initial assessment review, additional information from the proponents and commissioned studies, 115 days of public hearings in 26 communities, some delays for court rulings, and a lengthy period of analysis and writing, the Panel elaborated and applied this framework. In the last chapter of its 679-page final report, the Panel summarized its analysis, showing how it evaluated the impacts in each issue area for the null option (no project), for the project as filed, and for a range of further development and project expansion scenarios up to and beyond what would deliver 18.6 billion m^3 per year of gas pipeline throughput (Mackenzie Panel, 2009, chap.19). As well, the Panel determined, in each case, what the impacts would be with and without effective implementation of the Panel's 176 recommendations, how the various impacts might interact, positively and negatively, and what trade-offs would remain.

The Panel's overall conclusion was that the project could make a positive contribution to sustainability in the Mackenzie Valley but only if the proponents and governments implemented all of the Panel's recommendations (Mackenzie Panel, 2009, pp.613–615). Of the recommendations, the most significant and demanding ones were directed to the governments. These centred on anticipation and management of cumulative effects, especially through guiding the pace and scale of development, and on use of the revenues and other opportunities provided by the exploitation of non-renewable resources to make a transition to 'a more diverse, flexible and lasting basis for livelihoods in the region' (Mackenzie Panel, 2009, p.602).

The receiving governments rejected key aspects of the Panel's advice, particularly those requiring interventions in economic development to manage cumulative effects (Canada and the Northwest Territories, 2010). But by the time the Panel reported, controversial but effective new technology for exploiting shale gas deposits much closer to the main North American markets had led to sharply reduced natural gas prices, making the Mackenzie project economically unfeasible for the foreseeable future. Whether the project eventually proceeds and, if so, under what surrounding governance arrangements, remains to be seen.

Lessons and prospects

The language of sustainable development and sustainability did not become popular in Canadian policy pronouncements until the mid-1980s and had little effect on Canadian environmental assessment regimes until the 1990s. Despite some significant and illuminating applications, sustainability assessment is still not firmly entrenched in Canadian assessment law and practice. The most ambitious examples – cases involving formal assessment processes of some sort, with open public deliberations on major proposed undertakings, comparative evaluation of competing options, explicit attention to the interactions of effects on communities and biophysical systems, and special focus on long-term implications – began in Canada in the 1970s, but they have been special individual phenomena. Openings for these exceptional cases have been provided by the alignment of particular forces, typically including active public concern, multi-jurisdictional involvement, independent adjudicators (e.g. joint review panels), important new players and influences (e.g. recognition of Aboriginal rights, see Chapter 13), and/or widely recognized problems for which no established process seemed potentially adequate (e.g. regional growth management).

In this record and trajectory, sustainability assessment in Canada reflects a common path for innovations that challenge convention – needing to find openings where the prevailing formal and informal rules are weak, experimenting and learning from experience in different contexts, vulnerable to accusations of going too far, and likely to seem inefficient (if only because cutting a new trail is slower than following a well-trod one).

Arguably all seven of the Canadian cases surveyed here were successes and failures. The initial three were trail-blazing initiatives that combined significant achievements with the practical difficulties that typically face pioneers. The Berger inquiry set an international standard for fair, thorough, and ambitious public review. It raised public awareness of different perspectives on 'development' and it played a major role in winning serious attention to Aboriginal land claims. But it also persuaded Canadian governments never again to appoint a single, capable, independent-minded jurist to run a major assessment review. The Ontario timber management assessment helped overcome longstanding barriers to more farsighted, multi-stakeholder and multi-purpose approaches to forest management, but even those whose arguments prevailed found the process insufferable. The CRD's growth management strategy effort brought a new and much more promising approach to urban planning, but it too was slow work, and vulnerable to piecemeal weakening.

The four joint panel review cases were admirably successful in gradually raising the bar of demonstrated possibility and proper expectation in sustainability assessment; however, these gains have not yet been entrenched in conventional practice. Perhaps the panels' greatest accomplishments have been in developing more rigorous and defensible approaches to analysis, addressing cumulative effects and other strategic issues and strengthening attention to legacy effects. Most of the panels also reached conclusions that the relevant governments were willing to accept. But some of these same authorities, including the federal government, have been weakening environmental assessment law in the name of 'streamlining' decision making and have shown no inclination to entrench the contribution to sustainability test more firmly in the law. As a result, strong, sustainability-based assessments remain mostly limited to the exceptional big cases that go to panel review. Most formal assessment practice is still focused on mitigation. And despite the demonstrated strategic-level strengths of the sustainability assessments reviewed here, there is little sign yet of government inclination to extend requirements for open, sustainability-based assessment to the world of plans, programmes, and policies.

In summary, Canadian sustainability assessments have had a mixed record, considered in the light of the six effectiveness categories presented in Chapter 8. While there have been notable advances in applying the core principles, building analytical rigour, achieving substantive gains and learning from experience, progress has been far from smooth and the needed consistency of commitment, clarity of process, and efficiency of application are far from established.

The future is, as usual, uncertain. While the horizon includes a wide variety of attractive possibilities for further case applications of sustainability in Canada, there is no guarantee that an update of this chapter in five years' time will find many more completed examples to discuss, at least among the big initiatives that have been the focus here. Smaller scale, implicitly sustainability-based assessment activities of the sort noted at the beginning of this chapter, continue to proliferate widely. In the long run, as global and regional unsustainability effects become more pressing, demands for more rigorous and effective sustainability assessments at all scales, and in both formal and informal processes, are likely to increase. In the meantime, the eclectic set of Canadian sustainability-based assessments so far provides a promising foundation for further advances. Recognizing that no single process represents Canadian practice, Table 11.1 presents some general conclusions about Canadian practice based on the evaluation criteria set out in Chapter 8 of this book.

Table 11.1 Summary notes on the effectiveness of sustainability assessment in Canada

Framework criterion	Questions asked	Canadian perspective
Procedural effectiveness	*Have appropriate processes been followed that reflect institutional and professional standards and procedures?*	Practice varies widely. Some particular assessments have been exemplary in covering all steps and pushing the boundaries. Most regimes cover the basic procedural steps, but are weak in some key areas. Strategic-level assessments are typically ad hoc or done in a low-credibility policy-based process. Adequate monitoring is rare.
Substantive effectiveness	*In what ways, and to what extent, does sustainability assessment lead to changes in process, actions, or outcomes?*	Where applied, sustainability assessment has set a much higher test (positive contribution to sustainability rather than mitigation of adverse effects), has led to rejection of some major projects, and has had substantial effects on the nature of approved undertakings. Unfortunately, conventional practice in most jurisdictions addresses only a portion of the sustainability agenda.
Transactive effectiveness	*To what extent, and by whom, is the outcome of conducting sustainability assessment considered to be worth the time and cost involved?*	Some applications have been very lengthy, in part due to the role they have played in sectoral transitions that are rarely quick and tidy, and due to the use of big project assessments to address major strategic issues. Significantly greater efficiencies may depend on the introduction of linked strategic and project-level assessments.
Normative effectiveness	*In what ways, and to what extent, does the sustainability assessment satisfy the listed normative imperatives?*	The most advanced assessments adopt comprehensive sustainability-based criteria and specify them for the case and context, with consideration of interactive effects and trade-offs. This remains rare, however.
Pluralism	*How, and to what extent, are affected and concerned parties accommodated into and satisfied by the sustainability assessment process?*	Stakeholder engagement is generally well established in Canadian assessment processes, sometimes with intervenor funding. Major sustainability-based processes with public hearings are considerably more participative than the much more common smaller scale, mitigation-centred processes.
Knowledge and learning	*How, and to what extent, does the sustainability assessment process facilitate instrumental and conceptual learning?*	Sustainability assessments open a larger agenda, particularly concerning socio-economic/ecological interactions, long term/legacy effects, and broader alternatives. This facilitates more open public deliberation on desirable futures and how best to reach them. Participant learning about substantive issues and means of exerting influence has been evident. Institutional learning has been slowed by resistance to assessment results that challenge conventional assumptions and practices.

References

Berger TR (1977) *Northern Frontier, Northern Homeland: The Report of the Mackenzie Valley Pipeline Inquiry* (Ottawa: Supply and Services Canada).

British Columbia (BC) (1996) *Local Governmental Act*, part 25, Regional Growth Strategies. *Revised Statutes of British Columbia*, chapter 323. [Online] Available at: <http://www.bclaws.ca/EPLibraries/bclaws_new/document/ID/freeside/96323_29 > (accessed 8 June 2011).

Boyle M, Gibson RB, Curran D (2004) 'If not here, then perhaps not anywhere: urban growth management as a tool for sustainability planning in British Columbia's Capital Regional District', *Local Environment*, 9(1), 21–43.

Canada and the Northwest Territories (2010) *Governments of Canada and of the Northwest Territories Final Response to the Joint Review Panel Report for the Proposed Mackenzie Gas Project, November 2010.* [Online] Available at: <http://www.acee-ceaa.gc.ca/default.asp?lang=En&n=71B5E4CF-1> (accessed 12 February 2011).

Canadian Environmental Assessment Agency (CEAA) (2004) *News Release: Federal Environment Minister, Chair of the Mackenzie Valley Environmental Impact Review Board and Chair of the Inuvialuit Game Council Establish Joint Review Panel for the Mackenzie Gas Project*, Yellowknife, NWT, 18 August 2004.

Dacks G (1981) *A Choice of Futures: Politics in the Canadian North* (Toronto: Methuen).

Dunster JA, Gibson RB, Cook HA (1989) *Forestry and Assessment: Development of the Class Environmental Assessment for Timber Management in Ontario* (Toronto: Canadian Institute for Environmental Law and Policy).

Environmental Assessment Board (EAB) (1994) *Reasons for Decision and Decision: Class Environmental Assessment by the Ministry of Natural Resources for Timber Management on Crown Lands in Ontario* (Toronto: EAB).

Gamble DJ (1978) 'Berger Inquiry: an impact assessment process', *Science, 199(4332)* 3 March 1978, 946–951.

Gibson RB (2006) 'Sustainability assessment and conflict resolution: reaching agreement to proceed with the Voisey's Bay nickel mine', *Journal of Cleaner Production*, 14(3/4), 334–348.

Gibson RB (2011) 'Application of a contribution to sustainability test by the Joint Review Panel for the Canadian Mackenzie Gas Project', *Impact Assessment and Project Appraisal*, 29(3), 231–244.

Government of Canada (2011) *Canadian Environmental Assessment Act*, S.C. 1992, c. 37, consolidation of amendments to 28 April 2011. [Online] Available at: <http: //laws-lois.justice.gc.ca/eng/acts/C-15.2/index.html> (accessed 11 May 2011).

Government of Newfoundland and Labrador, Government of Canada, Innu Nation and Labrador Inuit Association (1997) *Memorandum of Understanding: Environmental Assessment of the Proposed Voisey's Bay Mining Development.* 31 January 1997; included as appendix to Voisey's Bay Panel (1999).

Government of Nova Scotia, Department of Environment and Labour (2007), *Whites Point Quarry: Minister's Decision.* 20 November 2007. [Online] Available at: <http://www.gov.ns.ca/nse/ea/whitespointquarry.asp> (accessed 8 June 2011).

Green, TL (1998) *Lasting Benefits from Beneath the Earth: Mining Nickel for Voisey's Bay in a Manner Compatible with the Requirements of Sustainable Development*, report for the environmental assessment hearings into the proposed Voisey's Bay nickel mine, prepared for Innu Nation.

IGC *et al.*, Inuvialuit Game Council, Mackenzie Valley Environmental Impact Review Board and the Minister of the Environment (Canada) (2004) *Environmental Impact Statement Terms of Reference for the Mackenzie Gas Project, August 2004.* [Online] Available at: <http://www.acee-ceaa.gc.ca/default.asp?lang=En&n=155701CE-1> (accessed 12 February 2011).

Inco Limited (2002) 'Inco Limited agrees on statement of principles with Province of Newfoundland and Labrador for development of Voisey's Bay deposits', news release, 11 June 2002.

Kemess North Joint Review Panel (2007) *Panel Report: Kemess North Copper-Gold Mine Project,* 17 September. [Online] Available at: <http://www.acee-ceaa.gc.ca/052/details-eng.cfm?pid=3394> (accessed 7 June 2011).

Mackenzie Gas Project Joint Review Panel (2005) 'Determination on sufficiency', 18 July 2005.

Mackenzie Gas Project Joint Review Panel (2009) *Foundation for a Sustainable Northern Future,* December 2009. [Online] Available at <http://www.ceaa-acee.gc.ca/default.asp?lang=En&n=71B5E4CF-1> (accessed 12 February 2011).

MVEIRB *et al.*, Mackenzie Valley Environmental Impact Review Board, Inuvialuit Game Council, Minister of the Environment (Canada) (2004) *Agreement for an Environmental Impact Review of the Mackenzie Gas Project.* August 2004. [Online] Available at: <http://www.ceaa-acee.gc.ca/default.asp?lang=En&n=71B5E4CF-1> (accessed 12 February 2011).

O'Faircheallaigh C (2006) 'Aborigines, mining companies and the state in contemporary Australia: a new political economy or "business as usual"?', *Australian Journal of Political Science,* 41(1), 1–22.

Page R (1986) *Northern Development: The Canadian Dilemma* (Toronto: McClelland and Stewart).

Robinson JB (1983) 'Pendulum policy: natural gas forecasts and Canadian energy policy, 1969–1981', *Canadian Journal of Political Science,* 16(2), 299–319.

Voisey's Bay Mine and Mill Environmental Assessment Panel (1997) *Environmental Impact Statement Guidelines for the Review of the Voisey's Bay Mine and Mill Undertaking.* [Online] Available at: <http://www.acee-ceaa.gc.ca/default.asp?lang=En&n=F06E8BD3-1> (accessed 13 February 2012).

Voisey's Bay Mine and Mill Environmental Assessment Panel (1999) *Report on the Proposed Voisey's Bay Mine and Mill Project.* March 1999. [Online] Available at: <http://www.ceaa-acee.gc.ca/default.asp?lang=En&n=0A571A1A-1&xml=0A571A1A-84CD-496B-969E-7CF9CBEA16AE&offset=2&toc=show> (accessed 7 June 2011).

Whites Point Quarry and Marine Terminal Joint Review Panel (2007) *Report on the Whites Point Quarry and Marine Terminal Project.* October 2007. [Online] Available at: <http://www.acee-ceaa.gc.ca/default.asp?lang=En&n=CC1784A9–1> (accessed 7 June 2011).

12 Sustainability assessment in South Africa

François Retief, North West University (Potchefstroom campus)

Introduction

South Africa with its exceptional natural beauty and cultural diversity is sometimes fittingly described as a 'world in one country' and 'the rainbow nation'. Although the country is world renowned for its achievements in the field of conservation and biodiversity management, it is also known for its turbulent political past marred by inequality and social injustice. It is a context where sustainability cannot be considered lightly because it deals with pressing short-term survival issues as well as longer-term concerns with quality of life. Therefore, considering the contribution and effectiveness of environmental assessment towards more sustainable decision making is particularly relevant in the case of South Africa.

South Africa represents a developing country with a mature and well-established environmental assessment system, even when compared to developed country systems (Lee and George, 2000; Wood, 2003). With over 4,000 EIAs conducted per year and more than 50 SEAs completed to date, the extent of practice is considered extensive (Retief *et al.*, 2007c; Retief *et al.*, 2011). Furthermore, the environmental assessment profession is particularly active as reflected by the over 1,000 members registered with the local South African chapter of the International Association for Impact Assessment (IAIAsa).

Similar to international experience there has been a strong focus in recent years on measuring the effectiveness of the environmental assessment system both through governmental reviews as well as independent research (DEAT, 2008a; 2008b; Retief, 2007a). At the outset it needs to be highlighted that the concept of 'sustainability assessment' *per se* does not exist within the South African context. However, dealing with sustainability has been implicit in the evolution of environmental assessment practice since its humble beginnings in the 1970s, and explicitly defined in the legal mandate for environmental assessment during the 1990s (Sowman *et al.*, 1995; Kidd and Retief, 2009). Therefore it is not surprising that the term has not been formally adopted since, by definition, South Africa theoretically has been doing sustainability assessment for at least the past two decades (Govender *et al.*, 2006).

Indeed the first main shift in the evolution of the environmental assessment system during the late 1970s and early 1980s was a move away from the focus on the biophysical environment or conservation to debates dealing with trade-offs between development and environment as well as the integration of the concept of sustainability with decision making (Kidd and Retief, 2009). Ultimately project-level EIA was legislated in 1997 and since then has gone through two extensive legal revisions in 2006 and 2010. Provisions for SEA were also introduced in the form of guidelines and enabling legislation in the 1990s, although no SEA regulations have been promulgated, and therefore strategic assessment is still conducted on a voluntary basis (DEAT, 2007). Throughout this chapter distinction is made between project-level assessment (i.e. EIA) and strategic-level assessment (i.e. SEA) because they represent two very different applications, albeit that both are considered synonymous within the understanding of sustainability assessment. Furthermore environmental assessment (EA) is used as an umbrella term that includes EIA and SEA.

As a point of departure it is necessary to explain the legal context for sustainability and its relation to the legal mandate for environmental assessment.[1] In its broadest sense the effectiveness of environmental assessment has to be considered against its ability to deliver on the so-called 'environmental right' enshrined in section 24 of the Constitution of the Republic of South Africa, 1996, which states that:

> Everyone has the right:
> (a) To an environment that is not harmful to their health or well being; and
> (b) To have the environment protected, for the benefit of present and future generations through reasonable legislative and other measures that
>   (i) Prevent pollution and ecological degradation;
>   (ii) Promote conservation; and
>   (iii) Secure ecologically sustainable development and use of natural resources while promoting justifiable economic and social development.

This right explicitly includes the concepts of intra- and intergenerational equity as well as sustainable development. However, it is not an absolute right and needs to be balanced against other rights in the constitution such as rights dealing with material well-being, i.e. access to housing, food, water, etc. To give effect to this environmental right a framework act called the National Environmental Management Act, 107 of 1998, (NEMA) was promulgated. The importance of this act for environmental assessment is that it provides definitions for the terms 'environment' and 'sustainable development', principles to be considered by all organs of state when taking decisions in terms of NEMA (which includes environmental assessment authorisations), and explicitly sets out objectives for environmental assessment. Table 12.1 provides a summary of the key definitions, principles and objectives contained in NEMA.

In a move towards implementation of sustainable development, government also developed a National Framework for Sustainable Development (NFSD), which sets out a conceptual understanding as well as a vision and key principles (DEAT, 2008c). As described in Figure 12.1 the NFSD supports a systems approach to sustainability, which moves away from the traditional concept

Table 12.1 The sustainability mandate for environmental assessment in South Africa as contained in NEMA

Selected NEMA Definitions

'Environment' means the surroundings within which humans exist and that are made up of

(i) the land, water and atmosphere of the earth;
(ii) micro-organisms, plant and animal life;
(iii) any part or combination of (i) and (ii) and the interrelationships among and between them; and
(iv) the physical, chemical, aesthetic and cultural properties and conditions of the foregoing that influence human health and wellbeing.

'Sustainable development' means the integration of social, economic and environmental factors into planning, implementation and decision making so as to ensure that development serves present and future generations'.

Selected NEMA Section 2 Principles (summarized)

Development must be socially, environmentally and economically sustainable.

Sustainable development requires the consideration of all relevant factors including the following: Avoidance and minimization of

(i) disturbance to ecosystem integrity and loss of biological diversity;
(ii) pollution and degradation of the environment;
(iii) disturbance of landscapes and sites that constitute the nation's cultural heritage;
(iv) waste;
(v) exploitation of non-renewable resources.

A risk-averse and cautious approach must be applied, which takes into account the limits of current knowledge about the consequences of decisions and actions.

NEMA Section 23 Objectives for environmental assessment

Promote the integration of the principles of environmental management set out in section 2 into the making of all decisions which may have a significant effect on the environment.

Identify, predict and evaluate the actual and potential impact on the environment, socio-economic conditions and cultural heritage, the risks and consequences and alternatives and options for mitigation of activities, with a view to minimizing negative impacts, maximizing benefits, and promoting compliance with the principles of environmental management set out in section 2.

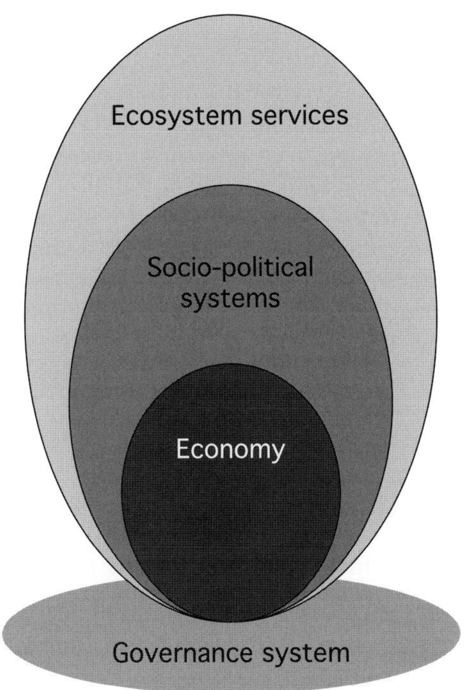

Figure 12.1

Systems approach to sustainability in South Africa (DEAT, 2008c)

of balancing the three pillars, biophysical environment, economic environment and social environment. It is within this context of sustainability provided by the constitution, NEMA and the NFSD that the effectiveness of EA in South Africa needs to be considered.

The following sections present a brief discussion against the effectiveness typology presented in Chapter 8. The different effectiveness 'elements' (procedural, transactive, normative and substantive) and 'influencing factors' (pluralism, knowledge and learning) are discussed separately before overall conclusions are made in the final section.

Procedural effectiveness

In the South African context procedural requirements are legally prescribed for EIA but not for SEA. However, the notion of 'ticking the box' towards procedural compliance has been central to some of the main criticisms of the assessment system. EIA practitioners, developers and government are increasingly frustrated and disillusioned with a system, which seems to have become procedurally overly complex and inefficient (O'Bearne and Boer, 2001; Macleod, 2006). It is argued that the legislative framework has led to the legalistic and mechanistic straight jacketing of assessment processes, transforming it into a lifeless and bureaucratic exercise, a move away from the need for flexibility and issues-driven approaches typical during the early years of EIA (Sowman *et al.*, 1996; Kidd and Retief, 2009).

Similarly, research into the quality of voluntary SEA processes also suggests an inability to incorporate flexibility into process design (Retief, 2007b). Instead, the understanding of SEA processes still considers the need to have to deal with clearly defined steps and phases within a 'rational' decision-making framework which need to be 'independent' and deliver 'objective' results. Therefore, in answering the question of whether environmental assessment achieves procedural effectiveness the answer would be that the procedural effectiveness discourse in South Africa is not concerned with compliance to specific procedural steps, but rather how to achieve flexibility in procedural design towards achieving context-specific sustainability objectives. It is also worth highlighting that the need for integrative thinking and flexibility in process design are also inherently included in the systems approach to sustainability described in the NFSD (DEAT, 2008c), illustrated in Figure 12.1.

Substantive effectiveness

Questions around substantive effectiveness are at the forefront of debates with research showing that since 2006 more than half of the 135 papers presented at the annual IAIAsa conferences dealt with issues of effectiveness (Retief, 2010). However, the only published empirical research has been in relation to SEA (Retief, 2007a). Measured against so-called 'direct' effectiveness indicators the research results show a high degree of ineffectiveness.[2] The main areas of weakness were the inability to influence the contents of plans and programmes as well as decision making in general. One of the main reasons identified for ineffectiveness was an inability to deal with the concept of sustainability, particularly in defining and measuring it, which led to methodological difficulties during scoping as well as follow-up monitoring and evaluation. Moreover, it also led in some cases to a lack of clarity on the exact purpose of the assessment. However, it would be incorrect to conclude that SEA has had no effect, because certain 'indirect effects' also emerged such as highlighting deficiencies and gaps in existing policy as well as examples where SEA facilitated capacity building and raised awareness of sustainability issues (Retief, 2007a; and Retief *et al.*, 2008). Moreover, SEAs have also contributed significantly to information generation and sharing.

When considering substantive effectiveness in a broader, more holistic sense (related to so-called 'indirect effects'), overall effects are wide ranging, positive and negative, across different scales and spheres of government (namely local, provincial/regional and national), as well as different areas or sectors of influence (such as the economy, politics, policy and legislation, etc.). The thinking around the substantive purpose of assessment has shifted from merely informing decision making (towards avoiding and minimizing impacts) to also aim to change decision-making cultures and normative principles underpinning decision making (Scott and Oelofse, 2005; Patel, 2009). In this regard the substantive purpose and normative elements should ideally be brought closer together. When considering the policy and legal mandate for EA, this has already happened in theory, but practitioners still struggle to give effect to or operationalize sustainability thinking towards more sustainable outcomes. It is only when sustainability becomes entrenched into normal EA processes, especially in a culture dominated by relatively inflexible procedural approaches discussed previously, that substantial gains might be expected to arise.

Normative effectiveness

As already indicated, environmental assessment is fully mandated in legislation to ensure that the normative sustainability aims are considered and incorporated in decision making affecting the environment. When comparing the NEMA principles and objectives highlighted in Table 12.1 as well as the NFSD approach in Figure 12.1 with the normative principles set out by Gibson in Chapter 1 it seems that implicitly and/or explicitly there is a strong correlation, especially in terms of recognising system complexities and the need to consider intra- and intergenerational equity, socio-ecological system integrity, maximizing of benefits and precaution.

However, the fact that sustainability has been defined and included as an objective of EA does not ensure success. The key characteristics of the South African definition of sustainability is that it is unashamedly anthropocentric and secondly that it fails to clearly explain how issues should be weighted and/or how trade-offs should be considered in decision making (Kidd, 2008). In a recent landmark constitutional court case the judge described the concept of sustainable development as a 'mediating principle' in EA which needs to reconcile and accommodate the three pillars of sustainable development, namely economic development, social development and environmental protection (Retief and Kotze, 2008).[3] From a deep ecology perspective this somewhat vague and clumsy description has been compared to 'shuffling deckchairs on the Titanic', because it does not recognise the intrinsic value of the environment nor does it deal with the fundamental flaws in the economic system which has been blamed for environmental destruction and extreme inequality (Pete, 2008). However, notwithstanding these challenges EA is already strongly positioned through its decision-making mandate to ask the right questions towards more sustainable outcomes. For example, environmental assessment should deal directly with issues of intra- and intergenerational equity, protection of livelihoods, resource maintenance and efficiency, precaution, etc. Initiatives to distil (directly and explicitly) context-specific so-called sustainability criteria for decision making from the legal mandate have already been explored by certain provincial environmental authorities (DEADP, 2010), and indications are that others are following. Thus the prospect for normative effectiveness in sustainability assessment in South Africa is promising, even though practice lags behind to date.

Transactive effectiveness

Questions on the cost implications of regulatory compliance in general and EIA in particular have been at the forefront of high-level political debate for some time. It has been estimated that regulatory compliance costs South African business a massive R796 billion per annum (6.5 per cent of total GDP for 2003), which is considered a substantial burden hindering national development targets (SBP, 2005; Macleod, 2006; Van Schalkwyk, 2006; Crookes and De Wit, 2002). In view of this increased concern over the cost of red tape to the economy, government has stated with reference to EIA that,

> Government is concerned about any delay, costs and associated impacts on economic growth and development. This is why we need to improve efficiency and effectiveness without compromising basic environmental rights and quality.
>
> (Van Schalkwyk, 2006, p.1)

The same message has been echoed across government, sometimes in less subtle ways. For example the minister of housing stated that:

> We cannot forever be held hostage by butterfly eggs that have been laid, because environmentalists would care about those things that are important for the preservation of the environment, while we sit around and wait for them to conclude the environmental studies.
>
> (Minister of Housing, Lindiwe Sisulu, as quoted by Macleod, 2006, p.12)

However, research on the efficiency of assessment processes shows that on average they are conducted in line with prescribed legal timeframes as well as international standards (DEAT, 2008b). In terms of direct cost, recent research also suggests that the average cost of EIA within South Africa is particularly low compared to international EIA systems (Retief and Chabalala, 2009). Similarly, research on the cost of SEA shows that different unrelated cases had distinct similarities in budget and took on average 11 months to complete (Kidd and Retief, 2009).

Overall the message within the South African context is that good practice EA does add value and saves time and money. This is because assessment asks, at an early stage in decision making, fundamental questions about the proposed action. These questions traditionally include issues around resource availability, efficiency in resource use, best practicable technologies, viable alternatives, etc. However, within the context of sustainability assessment questions around so-called indirect costs are also included. These indirect costs relate to loss of ecosystem services, impacts on quality of life, etc. It is recognized that although these costs are more difficult to calculate, and cover longer time frames, they are very important considerations when dealing with sustainability in EA (Retief, *et al.*, 2007d; Retief and Chabalala, 2009).

Pluralism

The depth and breadth of pluralism in South African society is represented by the diversity of cultures, races, socio-economic groupings, languages, etc. At the same time the country has some of the most liberal and extensive provisions for public participation, access to information and *locus standi* in the world. The opportunity to, in principle, have your voice heard has been central to the post 'apartheid' era democracy. Although in practice South Africa could not be described as a mature democracy, there are good examples where public participation and the right to access to information have, through the assessment process, influenced decisions on major developments related to mining and infrastructure (Glazewski, 2005; Kidd, 2008).

Furthermore, the definition of sustainable development in NEMA includes consideration of present and future generations. All these rights are explicitly entrenched in the constitution and have been given effect through extensive law reform and subsequent EA case law. However, very limited empirical research has been conducted to explore pluralism within the assessment system (Scott and Oelofse, 2005). Because of the complexity and diversity of South African society it can reasonably be expected that many different perspectives on the role and purpose of EA exists, beyond what is described in the legislation. Similarly, views within South African society will vary as to the effectiveness of the assessment process to consider the views of those involved. Anecdotal evidence suggests that affected role-players tend to be satisfied with the level of involvement if the final decision is in their favour and vice versa.

Knowledge and learning

An analysis of papers presented at IAIAsa conferences over the past decade shows that learning in EA happens in an incremental and/or muddled fashion and not according to a neat linear and/or logical progression, which is probably not unique to EA (Retief, 2010). There are proven instances of redesigning the wheel, but also instances of rapid growth in the complexity and range of debates. Therefore, attempts to explore and/or gain a better understanding of EA debates and learning need to recognize complexity and non-linear progression. The latter also suggest that tracing instrumental learning would be very difficult, and proving causality between EA and learning even more so. However, research shows that debates have shifted away from concerns with the quality and application of environmental assessment towards serious questions about effectiveness and the value that environmental assessment is adding and therefore the extent of instrumental learning achieved (Retief, 2010).

Conceptual learning is something that seems to have been at the centre of the evolution of environmental assessment in South Africa. As already indicated in the introduction, the concept of sustainability has been an early theme in the evolution of the assessment system. Over time principles such as equality, polluter pays, intra- and intergenerational equity, etc. have made their way into mainstream environmental law and subsequently the environmental assessment mandate. So in the case of South Africa it is difficult to determine to what extent the social context influenced conceptual learning in assessment, and to what extent assessment influenced conceptual learning within society. These two processes are probably interactive and mutually reinforcing.

However, learning happens against some form of knowledge. In recent years considerable work has gone into developing so-called 'sustainability science' in South Africa (Burns and Weaver, 2008). The 'sustainability science' research has focussed on means to define and conceptualize knowledge and learning complexity, especially in relation to environmental assessment (Burns *et al.*, 2006; Audouin and Hattingh, 2008). Of particular interest for South Africa is research conducted by Audouin and De Wet (2010) who identify different types of knowledge. The researchers argue that one of the key constraints towards introducing sustainability thinking into environmental assessment has been difficulties in engaging with 'value-based' and 'experiential' knowledge. Because assessment deals with very complex system thinking, traditional 'scientific' knowledge alone does not suffice. It is recommended that to deal with sustainability through assessment in the South African context requires an understanding of different knowledge types and ways of thinking (Audouin and Hattingh, 2008; Audouin and De Wet, 2010). The theory behind knowledge and learning related to sustainability and assessment seems to suggest the need for major mind-shifts. However, EA practice still seems to grapple with how to practically implement these ideas within a legalistic, procedurally driven EA system.

Summary and conclusions

This chapter reflects on the South African EA system against the effectiveness typology presented in Chapter 8. The result of the evaluation did not produce an effectiveness score or conclusive result, but rather a brief outline of the state of practice and lessons learned. In conclusion, South Africa presents a context where the EA system has an explicit mandate to incorporate sustainability into decision making. Although this sustainability journey started more than two decades ago, the environmental assessment profession has still not clearly refined or clarified exactly what sustainability means for EA decision making in practice. However, this somewhat vague destination has not prevented the environmental assessment community from persevering – which brings to mind the following discussion between Winnie the Pooh and Christopher Robin:

> 'Where are we going?' said Pooh, hurrying after him.
> 'Nowhere in particular' said Christopher Robin
> So they began going there.
> After they had walked for a while,
> Pooh asked 'Have we got there yet?'
> 'No' said Christopher Robin.
> So they kept on going!
>
> (Milne, 1928)

The question within the South African context is therefore not whether sustainability is worth pursuing or how the EA system should be changed or adapted to allow for sustainability assessment – but rather how can the existing and explicit mandate to pursue sustainability be operationalized in practice? Therefore, clearly the challenge lies in refining the questions that need to be asked, moving away from the traditional avoidance and minimization thinking towards long-term benefits and no net-loss outcomes. The formulation of sustainability decision criteria might be a necessary first step towards giving effect to the NEMA principles and objectives. As stated in the introduction, for South Africa, and many other developing countries, sustainability no longer only resonates with vague long-term utopian ideals, but deals with hard-hitting questions related to short-term survival and quality of life. The need for effective sustainability assessment has never been more acute.

At some risk of over simplification, Table 12.2 presents a summary of the results against the relevant framework criterion and questions.

Notes

1 South Africa has an extensive policy and legal framework dealing with sustainability. Although the focus of this chapter is on environmental assessment, it needs to be highlighted that numerous other sectoral legislation dealing with planning, water management, cultural heritage, etc. also include sustainability as a principle and/or objective.
2 For a more detailed description of the results and related methodology and indicators see Retief (2007a) and Retief (2007b).

Table 12.2 South African perspective on the effectiveness of sustainability assessment

Framework criterion	Questions asked	South African perspective
Procedural effectiveness	*Have appropriate processes been followed that reflect institutional and professional standards and procedures?*	To deal with sustainability there must be a move away from stipulated/prescribed steps towards more flexibility in process design. Procedural flexibility also facilitates flexibility and creativity in decision making.
Substantive effectiveness	*In what ways, and to what extent, does sustainability assessment lead to changes in process, actions, or outcomes?*	Research shows that SEA especially has not been effective in changing decisions or content of plans. However, EA has had a substantial so-called 'indirect' effect beyond the specific project or assessment, which needs to be recognized as part of the overall effectiveness debate.
Transactive effectiveness	*To what extent, and by whom, is the outcome of conducting sustainability assessment considered to be worth the time and cost involved?*	Assessment has come under severe criticism from politicians and developers for taking too long and costing too much. However, research suggests that both the cost and efficiency of EA are well within acceptable local and international standards.
Normative effectiveness	*In what ways, and to what extent, does the sustainability assessment satisfy the listed normative imperatives?*	EA has a strong sustainability mandate, including most of the listed normative imperatives (see Chapter 8). However, decision makers have only recently started to develop sustainability criteria in relation to these imperatives, so there is uncertainty as to how to give effect to them.
Pluralism	*How, and to what extent, are affected and concerned parties accommodated into and satisfied by the sustainability assessment process?*	Extensive provision is made for public participation, access to information and *locus standi*. Anecdotal evidence suggests that in general interested and affected parties are satisfied with opportunities provided for participation and consultation.
Knowledge and learning	*How, and to what extent, does the sustainability assessment process facilitate instrumental and conceptual learning?*	Learning by all role players involved in the assessment process happens in an incremental and muddled fashion which makes it difficult to trace types of learning and argue causality with EA. However, debates have shifted away from the application of EA to serious questions about added value and effectiveness. There is recognition that different types of knowledge such as 'value based' and 'experiential' knowledge are required to enhance the effectiveness of EA.

3 Surprisingly, the judge did not seem to consider the 'systems approach' propagated in the NFSD but rather the more dated 'three pillars' understanding.

References

Audouin M and Hattingh J (2008) 'Moving beyond modernism in environmental assessment and management'. In: Burns M and Weaver A (eds) *Exploring Sustainability Science – A Southern African Perspective* (Stellenbosch: African Sun Media), pp.205–242.

Audouin M and De Wet B (2010) *Applied Integrative Sustainability Thinking (AIST): An Introductory Guide to Incorporating Sustainability Thinking into Environmental Assessment and Management* (Pretoria: Council for Scientific and Industrial Research – CSIR).

Burns M, Audouin M, Weaver A (2006) 'Advancing sustainability science in South Africa', *South African Journal of Science*, 102, 379–384.

Burns M and Weaver A (2008) *Exploring Sustainability Science – A Southern African Perspective* (Stellenbosch: African Sun Media).

Crookes D and De Wit M (2002) 'Environmental economic valuation and its application in environmental assessment: an evaluation of the status quo with reference to South Africa', *Impact Assessment and Project Appraisal*, 20(2), 127–134.

DEADP (2010) 'Sustainability criteria for planning and EIA in South Africa', Paper presented at the annual IAIAsa conference (Pretoria, South Africa: DEAPD).

DEAT (2007) *Strategic Environmental Assessment Guideline – Integrated Environmental Management Guideline Series 4* (Pretoria: Department of Environmental Affairs and Tourism).

DEAT (2008a) *Capacity Audit and Needs Analysis Survey for Environmental Impact Assessment Administrators* (Pretoria: Department of Environmental Affairs and Tourism).

DEAT (2008b) *Draft Report: Review of the Effectiveness and Efficiency of the Environmental Impact Assessment (EIA) System in South Africa* (Pretoria: Department of Environmental Affairs and Tourism).

DEAT (2008c) *People-Planet-Prosperity, A National Framework for Sustainability in South Africa* (Pretoria: Department of Environmental Affairs and Tourism).

Glazewski J (2005) *Environmental Law in South Africa – Second Edition* (Durban: Lexis Nexis Butterworths).

Govender H, Hounsome R, Weaver A (2006) 'Sustainability assessment: dressing up SEA? – experiences from South Africa', *Journal of Environmental Assessment Policy and Management*, 8(3), 321–340.

Kidd M (2008) 'Removing the green-tinted spectacles: the three pillars of sustainable development in South African environmental law', *South African Journal of Environmental Law and Policy*, 15(1), 85–102.

Kidd M and Retief F (2009) 'Environmental assessment'. In: Strydom H and King N (eds) *Fuggle and Rabie's Environmental Management in South Africa* (Cape Town: Juta Publishing), pp.971–1047.

Lee N and George C (2000) *Environmental Assessment in Developing and Transitional Countries* (Chichester: John Wiley and Sons).

Macleod F (2006) 'Ministries aim to trash green laws', *Mail and Guardian*, 20 March 2006.

Milne AA (1928) *Winnie-the-Pooh* (London: Methuen and Co Limited).

O'Bearne S and Boer A (2001) 'Environmental assessment and sustainable development: hiding behind the facade of process'. Paper presented at the annual IAIAsa conference (White River, South Africa).

Patel Z (2009) 'Environmental justice in South Africa: tools and trade-offs', *Social Dynamics*, 35(1), 94–110.

Pete S (2008) 'Shuffling deckchairs on the Titanic? A critique of the assumptions inherent in the South African Fuel Retailers Case from the perspective of deep ecology', *South African Journal of Environmental Law and Policy*, 15(1), 103–126.

Retief F (2007a) 'Effectiveness of strategic environmental assessment (SEA) in South Africa', *Journal of Environmental Assessment Policy and Management*, 9(1), 83–101.

Retief F (2007b) 'A performance evaluation of strategic environmental assessment (SEA) processes within the South African context', *Environmental Impact Assessment Review*, 27, 84–100.

Retief F (2010) 'The evolution of environmental assessment debates – critical perspectives from South Africa', *Journal of Environmental Assessment Policy and Management*, 12(4), 1–23.

Retief F and Chabalala B (2009) 'The cost of Environmental Impact Assessment (EIA) in South Africa', *Journal of Environmental Assessment Policy and Management*, 11, 51–68.

Retief F and Kotze L (2008) 'The lion, ape and the donkey: cursory observations on the misinterpretation and misrepresentations of environmental impact assessment (EIA) in the chronicles of the Fuel Retailers', *South African Journal of Environmental Law and Policy*, 15, 138–155.

Retief F, Jones C, Jay S (2007c) 'The status and extent of Strategic Environmental Assessment (SEA) practice in South Africa – 1996–2003', *South African Geographic Journal*, 89(1), 44–54.

Retief F, Jones C, Jay S (2008) 'The emperor's new clothes – reflections on SEA practice in South Africa', *Environmental Impact Assessment Review*, 28, 504–5.

Retief F, Marshall R, Morrison-Saunders A (2007d) 'Avoiding extinction – proving the business case for environmental assessment (EA)', Paper presented at the annual IAIA conference (Seoul, South Korea).

Retief F, Welman C, Sandham L (2011) 'Performance of environmental impact assessment (EIA) screening in South Africa: a comparative analysis between the 1997 and 2006 EIA regimes', *South African Geographical Journal*, 93(2), 1–18.

SBP (2005) *Main Report: Counting the Cost of Red Tape for Business in South Africa.* (Johannesburg: Strategic Business Partnership).

Scott D and Oelofse C (2005) 'Social and environmental justice in South African cities: including "invisible stakeholders" in environmental assessment procedures', *Journal of Environmental Planning and Management*, 48(3), 445–467.

Sowman M, Fuggle R, Preston G (1995) 'A review of the evolution of environmental evaluation procedures in South Africa', *Environmental Impact Assessment Review*, 15, 45–67.

Van Schalkwyk M (2006) 'Environmental protection: quicker, simpler, better. A new EIA aystem for South Africa'. Keynote speech by Marthinus van Schalkwyk, Minister of the Department of Environmental Affairs and Tourism at the launch of the new EIA Regulations, April 2006 (Pretoria: Department of Environmental Affairs and Tourism).

Wood, C (2003) *Environmental Impact Assessment: A Comparative Review* (Harlow: Prentice Hall).

Part 4

Solutions?

13 Better engagement

Ciaran O'Faircheallaigh, Griffith University
Richard Howitt, Macquarie University

Introduction

Community engagement in various forms is widely supported as both strategy and process in securing improved environmental, social justice and sustainability outcomes from public decision making (Head, 2007; ICCM, 2010), and the importance of public participation in environmental decision making and engagement of affected communities in various forms of impact assessment has been widely recognised in a range of institutional settings. The 1992 Rio Declaration (Boer, 1995), the European Aarhus Convention (Hartley and Wood, 2005) and international bodies such as the International Association for Impact Assessment (www.iaia.org) and International Association for Public Participation (http://iap2.affiniscape.com/index.cfm) emphasise the value of public participation. In the specific context of sustainability assessment and the challenge created by pluralism (see Chapter 3), engagement strategies offer a means of integrating pluralism into sustainability assessment practice (see Chapter 8).

While this may appear to provide a way forward for sustainability assessment, the experience across impact assessment fields is that public participation is difficult and provides no certainty about the effectiveness of assessment, nor its outcomes. Doelle and Sinclair suggest that it is 'frequently . . . criticized as ineffective by participants, costly, and time consuming, by proponents, and inefficient by governments' (2006, p.186). Although public participation is a key theme in the impact assessment literature, there is little common ground about methods, standards, limits or goals for participation (O'Faircheallaigh, 2010). In the development studies field, Cooke and Kothari (2001) argue that many forms of participation and engagement have become compulsory in ways that constitute a new tyranny in many settings. Similarly, in environmental assessment, Nelson *et al.* (2008) ask just what sort of 'marriage' is anticipated by the various forms of public engagement undertaken for environmental purposes, while Leal (2007) interprets the re-politicising of participation as a tool in the neoliberal political agenda.

Clearly, community engagement in and of itself is no singular or simple solution. Fostering a critical understanding of what makes some forms of engagement better than others, and what continuing challenges arise from even the most appropriate and effective forms of engagement is critical to the development of an appropriate skills set for sustainability assessment. In considering what constitutes 'better', 'effective' or 'appropriate' engagement, it is necessary to recognise that engagement is of limited benefit to communities unless it provides them with a substantial capacity to shape decisions that affect their well being, and can only enhance the quality of decision making if the sustainability of agreed outcomes over time is assured.

For sustainability assessment practitioners pursuing procedural, substantive, normative and transactive effectiveness, strategies to engage public contributions and to facilitate high levels of stakeholder buy-in will be crucial. This chapter explores the idea of 'better engagement' in sustainability assessment in terms of the challenges in addressing pluralism, both as social diversity and as competing views of sustainable and just outcomes from assessment processes. Drawing on our extensive work in the social impact assessment of resources development on Indigenous land, we discuss Indigenous experience as an exemplar of the challenges of community engagement in the assessment process. While the marginalisation imposed on Indigenous groups in post-colonial settings is often extreme, making the issues of effective engagement very stark, it is not exceptional. For many local communities, ethnic, cultural and religious minorities, and specific social groups, such as youth, women, GLBTI (gay, lesbian, bi-sexual, trans-sexual and inter-sexual) and minority interest groups, the incapacity of dominant decision-making cultures to recognise and acknowledge their specific needs, concerns and rights parallels the lack of recognition of Indigenous peoples that we discuss here.

The chapter is structured around three basic questions about engagement:

1. How to engage (methods and process)
2. How to ensure that engagement involves a role in decision making (outcomes)
3. How to ensure that decisions are 'embedded' in ways that create sustainable outcomes.

Each of these dimensions involves significant challenges, in ascending order of magnitude. The chapter draws on extensive practical experience in social impact assessment and cross-cultural negotiation in the pluralist setting of the Australian resources sector, supplemented by examples from North America and the Asia Pacific.

Letting many voices be heard: methods for better engagement

In identifying basic principles, operating principles and guidelines for best practice in public participation in impact assessment, André *et al.* (2006, pp.2–3) highlight the importance of effective communication of information between the engaging parties and an accurate understanding by each of what the other is communicating. The information communicated will include:

- factual matters, such as how much land will be required for a project, what financial and other resources will be applied to support a proposed policy
- intentions and goals, including information on what each party wishes and intends to achieve
- principles and values, indicating issues on which compromise is not possible; how anticipated economic or social changes will be evaluated; and what values will be prioritised in evaluations.

In terms of sustainability outcomes, it is critical that voices representing affected economic, environmental, cultural and social values and interests are heard, accurately and fully, that none are excluded, and that their contribution is recorded to ensure that authorities and developers can be held accountable for their listening. It is easy to find examples of dramatic failures to achieve sustainability as a result of a failure by decision makers to listen to and heed the voices of affected people (see Box 13.1 on the Bougainville Copper Project). Such examples highlight the need to explore opportunities for success, and to identify both appropriate criteria for evaluating the adequacy and effectiveness of engagements, and the conceptual and methodological foundations for securing it. Better engagement, we suggest, not only secures more widely supported and sustainable outcomes in the first instance, but also provides stronger foundations for longer-term processes of engagement in implementation, review and adjustment that facilitate continuing progress towards sustainability and justice.

In our experience, at least three preconditions are essential for achieving effective communication (Agius *et al.*, 2002; O'Faircheallaigh, 2000). The first is a willingness by each party to listen to the other. In the context of projects affecting Indigenous peoples, this willingness has often been absent on the part of governments and companies. Decision makers have not accepted the need to engage with affected Indigenous people, as in the case of Bougainville, either denying the existence of a legitimate Indigenous interest in relevant outcomes, or believing that they themselves are able to identify and take account of any such interest. The second precondition is the physical ability to accurately communicate information, assuming that a willingness to listen is present. This represents a formidable requirement given that few non-Indigenous decision makers are fluent in Indigenous languages, and that in many settings, Indigenous people have limited capacity to comprehend non-Indigenous languages. The issue here is not just one of understanding the literal meaning of words, but of grasping the cultural context and world-view in which those words are understood and applied. Translation of words or concepts into what are regarded as their equivalent in another language does

Box 13.1 Unsustainable outcomes: project closure at the hands of those locked out of the process: Bougainville

In 1966–1967 an Australian affiliate of Rio Tinto, CRA Ltd, discovered what was to become one of the world's largest copper mines on the island of Bougainville, then part of the Australian colony of Papua and New Guinea. An Agreement to develop the project was negotiated between the Australian Administration and CRA Ltd. Bougainville villagers whose land would be required for the mine, for disposal of tailings, and for a port and township were not consulted and, despite the strenuous efforts of their Member of the (advisory) Legislative Assembly, Paul Lapun, were not allocated any share in the revenues that would flow from the mine. Some villagers opposed the development, and in one case riot police were used to quell their protests. Construction started in 1969, and was completed in 1972, the year before Papua New Guinea (PNG) achieved self-government. The project was highly profitable, and in 1974 the PNG government renegotiated the Agreement to give it a much larger slice of project revenues. However landowners continued to receive only a small share of benefits. In 1988–1989 an armed revolt by landowners, fuelled by massive environmental damage and internal conflict over distribution of limited benefits, closed the mine, and led to a military blockade by PNG and armed conflict between Bougainvilleans and PNG and among various factions on Bougainville. Thousands of people died, and the mine and associated infrastructure were destroyed. The mine remains closed today.

not mean that their meaning has been accurately conveyed. Box 13.2 illustrates these issues in the context of Native title negotiations in South Australia. Commitment to developing intercultural understanding is foundational to developing trust between parties, and confidence in process, which Doelle and Sinclair (2006) suggest are central to successful public participatory procedures.

Third, affected minority or marginalised interests must be able to interact and communicate internally in ways that are appropriate and effective in terms of their cultural and social norms. Positions, goals or values that need to be communicated to decision makers do not 'exist' or emerge automatically as soon as the need to articulate them arises. They must be developed and articulated through processes that incorporate the 'correct' people and do so in a way that properly reflects their cultural authority and their interests in the matter involved. In those Indigenous settings where communication is primarily oral and where kin and social relations are of central importance, this usually involves bringing people together to interact face-to-face, often on multiple occasions and over extended periods of time. Substantial resources are required to allow this to happen, given the remoteness and limited infrastructure of the regions in which many impact assessment projects engage with Indigenous people (O'Faircheallaigh, 2009; 2010).

Box 13.2 Using interpreters in South Australia

In 1999 the South Australian Government invited Aboriginal Native title claimants across the state to consider the possibility of developing a negotiated settlement of more than 20 separate Native title claims. The South Australian Aboriginal Legal Rights Movement (ALRM), then the representative body for all the state's Native title claim groups, set about engaging the affected groups to make a decision as to whether or not to pursue such negotiations.

The diverse claimant groups had not previously worked together at a whole-of-state scale, and the legal technicalities of the claim process had often led to decisions about claims being heavily influenced by non-Indigenous legal advisers. While the prospect of negotiating a comprehensive settlement of Aboriginal claims and concerns at the whole-of-state scale was exciting, developing a process that would ensure the claimants themselves were engaged as decision makers and were able to deliver free, prior and informed consent on any decisions to negotiate, or any proposed agreements to be settled, presented substantial challenges.

Led by Parry Agius, the *Narrunga* Chief Executive Officer of the ALRM Native Title Unit, the engagement process saw the formation of a new representative process (see Agius *et al.*, 2002; 2004; 2007) and adoption of some important and innovative practices. It also tackled the problems of cross-cultural engagement with government and industry parties from the mining, agricultural and fisheries sectors.

Perhaps most important was the use of interpreters in the discussions hosted by ALRM. In these large meetings of Aboriginal people from 23 Aboriginal nations, most people spoke English, but many were unused to public speaking, and unfamiliar with many of the challenges raised by such sensitive political negotiations. Although it would have been quicker and less expensive to conduct all proceedings in English and rely on informal interpreting by participants for those not fluent in English, use of interpreters not only focused the process on an Aboriginal-centric orientation to the issues, but also slowed proceedings down in a way that gave people time to think and to test the ideas and proposals that were being developed. Requiring all speeches to be interpreted also supported contributions by many people who were less confident in public speaking, and gave voice to some whose poor English skills might have otherwise excluded them from the process.

For government and industry participants, stopping to have their words presented in Aboriginal language to an Aboriginal audience was an unfamiliar and unsettling experience. At one point, the state's Attorney-General expressed frustration at the slowness of the process, preferring to move quickly to a decision, rather than deliberate on the content of a speech. The indomitable interpreter simply pulled him into line by saying that everyone in the room realised he was a very important man and they wanted to really understand every word he said – so he should simply stop every few sentences and allow her to ensure he was being understood by everyone present.

In another salutary lesson, key Native title claimants reminded people that their language was central to their identity and it was important to insist that governments learn to listen to what was said about their country, their culture and their futures in their own words – not the words of the invader. During this particular meeting, the interpreters translated two distinct and important concepts ('negotiation' and 'agreement') with the same *Antakarinya* word, and the planned agenda was suspended for nearly four hours while participants explored the best way to understand the ideas behind the English terms to ensure that any consent to negotiate or to reach an agreement was fully informed and debated.

In the case of resource industries' engagements with Indigenous Australians, formidable barriers exist to achieving these preconditions to effective communication, including racism; a tendency towards excessive emphasis on formal structures and processes in environmental and social impact assessments; a failure to acknowledge the importance of Indigenous knowledge in achieving accurate and comprehensive assessments; and skewed allocation of resources between assessment of environmental and Indigenous impacts, and among the various interests involved in decisions that affect Indigenous peoples (Carter and Hill, 2007; Collier and Scott, 2009; O'Faircheallaigh, 2002). On the other hand effective communication and articulation of Indigenous interests is not impossible to achieve, and a substantial number of successes in this regard are documented in the literature (Agius *et al.*, 2002; O'Faircheallaigh, 2000; Kimberley Land Council, 2010a).

Influencing decisions

It is one thing to be listened to, another to have a capacity to influence and shape decisions. The wider literature on public participation in decision making reflects this reality, noting that governments and corporations may have no intention of sharing their control over decisions but rather only engage with Indigenous or other groups in order to obtain from them what is required for decision making or in an attempt to nullify opposition to decisions that have already been made (Arnstein, 1969; O'Faircheallaigh, 2010). The need for participation to include a real say in decision making is widely recognised as an essential part of good practice and as a value that should underlie community engagement (André *et al.*, 2006; Collier and Scott, 2009). It is critical to consider what legal and political mechanisms are available to achieve this outcome in practice. In some cases Indigenous peoples may have a constitutional or legislative right to be involved in decisions regarding development on their traditional estates. In Australia's Northern Territory, for example, owners of Aboriginal freehold land are granted an effective veto over new developments under the *Aboriginal Land Rights (Northern Territory) Act 1976*. This allows them to determine whether or not development should occur, and places them in a strong position to negotiate regarding the specific shape of any development they are prepared to approve. In the Canadian north, comprehensive land claim settlements concluded during recent decades typically give Indigenous groups complete control over development on key areas of land (for instance areas surrounding communities, key wildlife breeding grounds), and a structured role in managing commercial activity throughout the land settlement area. There is also a constitutional duty on the Crown in Canada to consult with Indigenous peoples regarding developments that affect them, though the extent to which this implies Indigenous influence over decision making is hotly contested.

Indigenous people can and do push for an active role in decisions that affect them in situations where they have limited or no legal rights to control what occurs on their ancestral lands. It is rare for interests that hold power to yield gracefully to groups pushing for a share of it, and Indigenous peoples have had to use a variety of approaches to try and insist on active and substantial participation in decision making. These include the use of regulatory mechanisms provided for example by environmental impact legislation; litigation, based for instance on mining legislation or on laws related to public participation in government decision making (including administrative appeals and freedom of information legislation); media campaigns designed to generate popular pressure on government and corporate decision makers; and direct action to delay or disrupt development activities. Success is most likely to be achieved where a number of these approaches, or all of them, are used in a coordinated fashion. For example the Innu and Inuit people affected by the proposed Voisey's Bay mine in Labrador, Canada, were initially excluded from decision making in relation to the project by the developer, Inco, and the Newfoundland government. The Innu and Inuit used the environmental review process to press for project approval conditions that would require their agreement before the mine could proceed (see also Chapter 11); engaged in litigation to try and prevent Newfoundland from issuing licences for the project; and occupied the mine site when the developer refused to engage with them and again when their court action was initially unsuccessful. The Innu and Inuit eventually achieved a major say over key decisions in relation to the project, including the fundamental question of its physical scale (with production levels less than a third of those initially planned by the developer), and the environmental conditions that would apply to the mine (Gibson, 2006; O'Faircheallaigh, 2007).

Conflict does not always surround the determination of Indigenous groups to be involved in decision making. For example, in 2006 the Government of Western Australia established, as a matter of policy, that development of Liquefied Natural Gas (LNG) processing facilities would only occur in the Kimberley region with the informed consent of affected Traditional Owners. It funded the regional Aboriginal land organisation, the Kimberley Land Council (KLC), to establish a process that would give concrete expression to this commitment, with the result that at least for a period Aboriginal Traditional Owners exercised a fundamental measure of control over development (see Box 13.3 for a summary of the process involved). However, illustrating the vulnerability of Indigenous interests, especially where they cannot draw on rights enshrined in legislation (see also Chapter 6), after an election in 2008 a new state government changed the policy on Indigenous informed consent, selected its preferred site for the LNG processing facility, and threatened to compulsorily acquire the land involved if Traditional Owners did not consent to its development.

Box 13.3 A cooperative approach to Indigenous participation in government decision making: the Kimberley LNG site selection process

In 2006 the (Labor) Government of Western Australia decided that, rather than have individual oil companies identify and develop their own sites for processing offshore natural gas along the Kimberley coast, the Government would seek a single location for an industrial precinct where processing of all gas would occur (an 'LNG Precinct'). The Government indicated that, as a matter of policy, an LNG Precinct would only be established in the Kimberley with the 'fully informed consent' of the Traditional Owners of any proposed site. In 2007 the State established a Northern Development Taskforce (NDT) to conduct the site selection process for an LNG Precinct, and agreed to fund the KLC to establish a Traditional Owner consultation and decision-making process that would give practical effect to the State's requirement for Indigenous informed consent. The Kimberley Land Council (KLC) established a Traditional Owner Task Force (TOTF) comprising representatives of all coastal native title groups and senior Kimberley 'cultural bosses' to establish and oversee an appropriate process. Consistent with traditional decision-making practices, the TOTF would not make decisions about whether to agree to the locating of a LNG hub in the Kimberley or on the traditional land and sea country of particular groups. Such decisions would need to go back to the whole native title-claim groups for areas being considered as suitable locations. The TOTF would make decisions about the process(es) of consultation about LNG development; ensure the integrity of the consultation and information delivered by the TOTF and the KLC project team members; and act as a conduit for the flow of information to and from the larger native title-claim group membership. It would also provide a mechanism through which the native title groups would support each other, whatever decision individual groups made about LNG development on their traditional country.

TOTF and native title group meetings occurred between April and September 2008. The TOTF engaged in exchanges with government, non-government and industry visitors all of whom presented information concerning the proposed LNG development and answered questions raised during the meetings. At each TOTF meeting an agenda was set, minutes were recorded, key issues and tasks to be undertaken were highlighted and questions unanswered or requiring further elaboration and detail were noted. These records formed the basis for preparing TOTF newsletters that were presented at the following meeting as a record of the meeting as well as the basis for discussion within families and the wider native title-claim groups.

Between July and August 2008 the KLC met with relevant native title-claim groups to determine which of the remaining 11 locations chosen for further consideration by the NDT, from 42 original possible locations, could remain in the site selection process. Traditional Owners for these proposed locations participated in scientific and engineering studies in collaboration with the NDT. As the process unfolded a number of Traditional Owner groups withdrew their land and sea country from consideration as potential sites, though these decisions were not made public until September 2009. Owners withdrew sites in some cases because multiple potential sites existed in their land and sea country and they only wished a single site to be considered. In other cases they withdrew sites because of their serious concerns about the potential impact of a Precinct on the environment and on their cultural and economic lives. The NDT site selection processes also removed some of these same sites from consideration due to environmental and/or technical considerations. In September 2008 the KLC formally announced that the Traditional Owners of seven potential locations had removed them from further consideration, with four others remaining under consideration.

The process of information provision and the ability of Traditional Owners to decide which sites would and would not continue to be considered represents a significant achievement in terms of giving Indigenous people real control over development on their traditional lands.

On 15 October 2008 the newly elected Government of Western Australia announced its preferred site for the location LNG Precinct and stated that the TOTF site-selection process would not be continued. While the preferred site was one of the four potential sites not removed from consideration, this decision in effect meant that Traditional Owners were not able to make their own decisions about whether these sites should be removed from consideration or continue to be assessed (Kimberley Land Council, 2010b).

Note: Extensive documentation is available regarding the Site Selection Process and subsequent developments in relation to the Kimberley LNG Precinct. Relevant information can be found on the websites of the Kimberley Land Council (http://klc.org.au/gas/) and the Western Australia Department of State Development (http://www.dsd.wa.gov.au/7909.aspx).

Making engagement work: embedding solutions in sustainable structures

Despite significant progress in the capacity of governments and development proponents to consult with and foster participation by affected communities in project-based assessment activities, engagement across difficult boundaries remains more an abstract ideal than a common practice. The challenge is to embed solutions that accommodate diversity and evolving social and cultural relationships in adaptable and accountable structures (Jacobs & Mulvihill, 1995). In many cases ongoing arrangements reflect existing patterns of power and privilege, with state agencies and corporate entities exercising decisive influence and control. Early exceptions to this general pattern offered a level of accountability to local monitoring of environmental performance (e.g. Alaska's Red Dog Mine, see Case, 1996; ICME, 1999) or a level of autonomy for community development activities (e.g. Howitt, 1995). More recently, negotiated agreements have drawn on discourses of Indigenous rights, sustainability and social justice to secure stronger and more sustainable outcomes (O'Faircheallaigh, 1999; 2004; 2007; 2009; O'Faircheallaigh &

Corbett, 2005). In the resource sector, negotiated agreements have become almost commonplace as the outcome of mining companies' engagements with Indigenous peoples (e.g. ICCM, 2010). But the priorities of companies and state agencies, rather than pursuit of sustainable outcomes that incorporate indigenous interests throughout project life, continues to frame most thinking about agreement-making. As a result, the best that advocates of sustainability can hope to secure is some form of compromise towards sustainability principles as a basis for initial decisions on project development. In situations of unequal power, deeply embedded racism or entrenched disadvantage, institutional responses to the need for just and sustainable outcomes for local and Indigenous stakeholders affected by resource projects are required that do more than simply draw them into participatory but unsustainable arrangements.

Folke *et al.* (2002, p.15) refer to 'complex adaptive systems' as the basis for building sustainable and resilient institutions capable of responding to rapidly changing circumstances. They suggest that four characteristics are critical for developing such adaptive systems:

- learning to live with change and uncertainty
- nurturing diversity for resilience
- combining different types of knowledge for learning
- creating opportunity for self-organization towards social-ecological sustainability.

In engaging with previously marginalised or excluded stakeholders, decision makers often seek ways to minimise their exposure to change and uncertainty – externalising such 'risks' onto others – rather than learning to live with and adapt to changing circumstances (Schipper & Pelling, 2006). Many state and corporate negotiators continue to respond to social and cultural difference in traditional thinking modes that insist on demonstrated progress towards the norms of dominant cultural and economic values, finding it difficult to accommodate different types of knowledge in developing new institutional arrangements (Castro and Nielsen, 2001). Continued adherence to centralised, state-defined criteria insists on negotiated arrangements that conform to old patterns of decision making rather than focusing on performance that is responsive to goals focused on sustainability and justice (O'Faircheallaigh & Corbett, 2005). In such circumstances, securing a 'place at the table' is not just a chance to speak out but also a small step towards challenging these old patterns and embedding sustainability solutions in institutional responses to sustainability assessment (Trebeck, 2007; Muller, 2008; Koivurova, 2010).

In many situations, the benefits of that initial challenge to old patterns of exclusion and exploitation are lost as momentum shifts away from consultative engagement and towards project implementation. Shifting from pre-development engagement to engaged institutions and structures that pursue sustainability outcomes remains a major challenge for many state and corporate actors. They assume the privileges of power all too easily in their engagement with diverse public interests, particularly where engagements occur across cultural differences. Demonstrating that sustainability assessment offers a practical and realistic way to address the challenges of pluralism, and embedding sustainability as a key criterion in assessing and managing major projects, is not only central to the credibility and effectiveness of sustainability assessment as a means of shaping the future (Chapter 2). It is also central to developing agreements that are themselves sustainable, agreements that can succeed across generational and contextual changes. There are several key dimensions to embedding sustainability solutions in project-based activities. Solutions have to work across a range of spatial and temporal scales (Chapter 4 & 5), be enforceable and accountable across jurisdictions (Chapter 6) and be inclusive of diverse stakeholders and interests (Chapter 7).

The implications of major projects across spatial and temporal scales are typically represented in ways that simplify complex socio-ecological systems to justify maximising resource flows into relatively short-term market-based processes (Folke *et al.*, 2002). Shifting decision making on such projects towards accountability for their sustainability implications requires a move away from what might be glossed as a politics of success towards a politics of responsibility. Both detractors and supporters of resource projects typically value costs, benefits, risks and opportunities differently. In the pluralist settings of contemporary discourses of both sustainability and globalisation, the complex politics involved in negotiating agreement about projects is not generally reducible to a choice between binary oppositions, even though decisions are often politicised as a stark choice between 'success' and 'failure'. Rather, the challenge is to negotiate what is referred to as 'wicked complexity' (Allen & Gould, 1986; Paconowsky, 1995; Lachapelle *et al.*, 2003), which involves negotiating sustainable relationships between stakeholders across temporal, spatial and value differences. As an exemplar of more general notions of such 'wickedness', Howitt (in press) characterises the context of development decisions in remote Indigenous territories in the following terms:

- Indigenous people, parliaments, state agencies, Indigenous organisations and others often disagree on basic goals
- mutual trust, respect and recognition have been deeply eroded
- the burden of administrative and procedural responsibilities and inflexible funding, reporting and delivery structures have been institutionalised and ossified
- the capacity to listen to Indigenous people and their own analyses of their own circumstances has been severely limited over a long period of time
- the development of institutions of governance and livelihood has not been oriented towards solutions.

In this context, engagement to secure sustainable and resilient resolution of wickedly complex social relations within socio-ecological connections must address all the criteria Folke *et al.* (2002) identify and embed flexible solutions in institutional outcomes. Sustainability assessment focused on the specific contribution of a particular project will not automatically respond to the needs and concerns of Indigenous advocates or environmentalists. Doing so will often require consideration of information that project advocates consider irrelevant or beyond the scope of their inquiries and a source of unnecessary and unacceptable risk. Yet sustainable solutions rely on buy-in from all those affected. This, in turn, requires a shift away from narrow, project-focused thinking, towards more broadly contextualised engagement. It requires serious attention to the emotional and procedural concerns of all parties rather than just the substantive issues targeted for resolution.

Our discussion has already identified some of the challenges that arise in securing effective engagement of communities, governments and project advocates in project assessment and negotiation processes. In the wickedly complex circumstances of the real world, however, engagement needs to extend beyond this phase into project management and decommissioning. Proponents of a particular development project typically focus on achieving a specific and limited range of substantive outcomes. Extending this focus towards more inclusive, resilient and sustainable systems across and beyond the life of a project shifts thinking out of a binary mode and impels engagement with social and ecological diversity. For state and corporate proponents of a project, every contributing decision, every constituent element of process, is linked to a simplified binary decision (yes/no). In contrast, for others drawn into engagement processes because they are somehow affected by a development proposal (or face or perceive risks arising from it), the landscape is not reducible to a set of binary choices or linear sequences leading to them. The relational setting that characterises complex webs of relatedness in social, biophysical and cosmological connections demands consideration of pluralism not simply in terms of diverse social interests, but also in terms of multiple methods for investigating available options and overlapping and contested jurisdictions for governance and regulation. There is no single solution to this challenge. As Ostrom & Cox put it:

> Addressing this complexity must in turn overcome historical academic divisions between ecology, engineering and the social sciences, the tendencies of social scientists to build simplified models of complex systems in order to derive ideal types of governance, and an overreliance on a limited set of research methods to study social and environmental systems.
>
> (Ostrom & Cox, 2010, p.451).

Natcher *et al.* (2005) argue that rather than simply achieving agreement, it is necessary to think about the sociality of proposed outcomes. This reinforces the need for a well-contextualised understanding of adaptive institutions in cross-cultural engagement (Jacobs & Mulvihill, 1995) and pluralistic, polycentric solutions rather than pre-determined panaceas (Ostrom & Cox, 2010). The literature on co-management of natural resource systems across culturally diverse stakeholders – a setting in which pluralism and wicked complexity are common – emphasises the links between method, process and structure. That is, more just and sustainable outcomes require the development of sustainable relationships between people as the basis for achieving both short-term project approvals or specific management targets and longer-term sustainability in complex intercultural systems (Natcher *et al.*, 2005; Rodriguez-Isquierdo *et al.*, 2010).

To secure viable structural and institutional responses that embed sustainability thinking in public participation and engagement processes undertaken as part of sustainability assessment, the foundational role of relationships needs to be understood, acknowledged and valued by regulatory authorities, developers and community stakeholders. If the goal is sustainability in some form, rather than simply approval of a development proposal that is *a priori* defined as desirable and sustainable by its supporters, the cultural, biophysical and cosmological dimensions of sustainability that are accepted and used by affected communities need to be reflected in both process and outcomes. Institutions intended to monitor sustainability outcomes, for example, must not only address the complexity of the current situation, but must also be responsive and adaptable to evolving circumstances and changing relationships over time, to the emergence of new sorts of connections across space and time, and to changing understanding of the implications of particular decisions and situations. Key characteristics of such institutions include an ability to operate at and link across geographical and temporal scales to secure effective place-based management (Brondizio *et al.*, 2009). In doing so new institutions must:

- integrate both predictions and critical reflection across systems (Young *et al.*, 2006)
- integrate research, communication and capacity-development into basic institutional procedures and structures (Leemans *et al.*, 2009)
- integrate both biophysical and socio-political elements into governance (Kotchen & Young, 2007)
- accommodate plural interests, perspectives and values (Howitt and Suchet-Pearson, 2003; 2006).

In pursing adaptive performance, such institutions will also be embedded within transparent and accountable systems of governance and planning, and watchful for maladaptive developments (Adger and Barnet, 2009).

Conclusions

Political decision making about development choices produces winners and losers. Perhaps the most important question, then, is just what sorts of engagement are most likely to generate an outcome that embodies sustainability principles and is sufficiently consensual to 'stick', even if not universally supported? We have argued that engagement must pay careful attention to the requirements for 'real' communication across cultures, so that participants understand each other fully and correctly, and be based on a realignment of political power so that all participants have a capacity to influence outcomes. Referring to the experience of Indigenous peoples as exemplary of the concerns facing a wide range of community interests in sustainability assessment, we have explored ways in which the effectiveness criteria discussed in Chapter 8 might be used to explore engagement processes. Using these criteria, procedural and normative elements of the impact assessment process can be re-contextualised as central to engaging diverse communities of interest to improve how sustainability is integrated transparently and fairly into decision making. Project-based outcomes, however, must also be embedded in both relationships and institutions in ways that allow them to be sustained over the long term. At the project scale, negotiation-based approaches combined with sustainability assessments offer a starting point. Scaling up from the project scale to whole-of-government, intergovernmental and global sustainability orientations, however, requires substantial rethinking of engagement. As our discussion of Indigenous experience demonstrates, simplistic democratic responses that rely on representation, hierarchy and tolerance by majority populations risks reinforcing national and global-scale patterns of exclusion and imposing unsustainable poverty and injustice on some people and places. The rapidity of regime change, political redirection and the impacts of natural disasters across many jurisdictions provide a salutary reminder that even the most enduring facilities and institutions – including well-resourced national governments and well-regulated nuclear power plants – face changing contexts that require adaptive responses.

Structural solutions to the challenges of project, process and social governance must not only secure buy-in through initial engagement and negotiation, but must continue to build, challenge and reinvent consensus over time (see Chapter 4), between groups (see Chapter 7) and jurisdictions (Chapter 6) and across geographical scales (see Chapter 5). For sustainability assessment, then, the challenge is not simply to deal with winners and losers. It is to recognise that, in sustainability terms, winning and losing is a constant and ongoing affair, in which the processes by which people in their various communities and guises need to engage, and to be engaged, in processes that constantly reconsider how to judge the sustainability, fairness and costs of arrangements put in place to govern human activities.

References

Adger WN and Barnett J (2009) 'Four reasons for concern about adaptation to climate change', *Environment and Planning A*, 41(12), 2800–2805.

Agius P, Davies J, Howitt R, Johns L (2002) 'Negotiating comprehensive settlement of native title issues: building a new scale of justice in South Australia', *Land, Rights, Laws: Issues of Native Title*, 2(20), 1–12. [Online] Available at: <http://www.aiatsis.gov.au/ntru/docs/publications/issues/ip02v2n20.pdf> (accessed 24 November 2011).

Agius P, Davies J, Howitt R, Jarvis S, Williams R (2004) 'Comprehensive Native title negotiations in South Australia'. In: Langton M, Teehan M, Palmer L, Shain K (eds) *Honour Among Nations? Treaties and Agreements with Indigenous People* (Melbourne: Melbourne University Press) 203–219.

Agius P, Jenkin T, Jarvis S, Howitt R, Williams R (2007). '(Re)asserting Indigenous rights and jurisdictions within a politics of place: transformative nature of Native title negotiations in South Australia', *Geographical Research*, 45(2), 194–202.

Allen GM and Gould EMJ (1986) 'Complexity, wickedness and public forests', *Journal of Forestry*, 84(4), 20–24.

André P, Enserink B, O'Connor D, Croal P (2006) *Public Participation International Best Practice Principles. IAIA Special Publication No 4* (Fargo: International Association for Impact Assessment). [Online] Available at: <www.iaia.org/public documents/special-publications/SP4%20web.pdf> (accessed 1 August 2011).

Arnstein S (1969) 'A ladder of citizen participation', *Journal of the American Institute of Planners*, 35(4), 216–224.

Boer B (1995) 'The globalisation of environmental law: the role of the United Nations', *Melbourne University Law Review*, 20(1), 101–125.

Brondizio ES, Ostrom E, Young OR (2009) 'Connectivity and the governance of multilevel social-ecological systems: the role of social capital', *Annual Review of Environment and Resources*, 34(1), 253–278.

Carter JL and Hill GJE (2007) 'Critiquing environmental management in indigenous Australia: two case studies', *Area*, 39(1), 43–54

Case D (1996) 'The Alaska experience: in a twinkling – the Alaska Native Claims Settlement Act and agreements relating to the use and development of land'. In: Meyers GD (ed.) *The Way Forward: Collaboration and Cooperation 'In Country'* (Perth: National Native Title Tribunal), pp.102–124.

Castro AP and Nielsen E (2001) 'Indigenous people and co-management: implications for conflict management', *Environmental Science & Policy*, 4(4–5), 229–239.

Collier MJ and Scott M (2009) 'Conflicting rationalities, knowledge and values in scarred landscapes', *Journal of Rural Studies*, 25(3), 267–277.

Cooke B, and Kothari U (eds) (2001) *Participation: The New Tyranny?* (London and New York: Zed Books).

Doelle, M and Sinclair AJ (2006) 'Time for a new approach to public participation in EA: promoting cooperation and consensus for sustainability', *Environmental Impact Assessment Review* 26(2), 182–205.

Folke C, Carpenter S, Elmqvist T, Gunderson L, Holling CS, Walker B, Bengtsson J, Berkes F, Colding J, Danell K, Falkenmark M, Gordon L, Kasperson R, Kautsky N, Kinzig A, Levin S, Maler K, Moberg F, Ohlsson L, Olsson P, Ostrom E, Reid W, Rockstrom J, Svenije H, Svendin U (2002) *Resilience and Sustainable Development: Building Adaptive Capacity in a World of Transformations. Series on Science for Sustainable Development No. 3* (Stockholm, International Council for Science). [Online] Available at: <http://www.sou.gov.se/mvb/pdf/resiliens.pdf> (accessed 24 November 2011).

Gibson RB (2006) 'Sustainability assessment and conflict resolution: reaching agreement to proceed with the Voisey's Bay nickel mine', *Journal of Cleaner Production*, 14(3–4), 334–348.

Hartley N and Wood C (2005) 'Public participation in environmental impact assessment – implementing the Aarhus Convention', *Environmental Impact Assessment Review*, 45(4), 319–340.

Head, BW (2007) 'Community engagement: participation on whose terms?', *Australian Journal of Political Science* 42(3): 441–454.

Howitt R (1995) *Developmentalism, Impact Assessment and Aborigines: Rethinking Regional Narratives at Weipa* (Darwin: North Australia Research Unit).

Howitt R (in press) 'Sustainable indigenous futures in remote Indigenous areas: relationships, processes and failed state approaches', *GeoJournal* 75: Online First [DOI: 10.1007/s10708–10010–19377–10703].

Howitt R and Suchet-Pearson S (2003) 'Ontological pluralism in contested cultural landscapes'. In: Anderson K, Domosh M, Pile S, Thrift N (eds) *Handbook of Cultural Geography* (London: Sage), pp.557–569.

Howitt R and Suchet-Pearson S (2006) 'Rethinking the building blocks: ontological pluralism and the idea of "management"', *Geografiska Annaler: Ser B, Human Geography*, 88(3), 323–335.

ICCM (International Council on Mining and Metals) (2010) *Indigenous Peoples and Mining: Good Practice Guide* (London: ICCM).

ICME (International Council on Metals and the Environment) (1999) *Mining and Indigenous Peoples: Case Studies* (Ontario, Canada: ICME).

Jacobs P and Mulvihill P (1995) 'Ancient lands: new perspectives. Towards multi-cultural literacy in landscape management', *Landscape and Urban Planning*, 32(1), 7–17.

Kimberley Land Council (2010a) *Kimberley LNG Precinct Strategic Assessment Indigenous Impacts Report: Six Volumes*. [Online] Available at: <http://klc.org.au/2010/12/09/james-price-point-indigenous-impacts-report-released/> (accessed 23 March 2011).

Kimberley Land Council (2010b) *Kimberley LNG Precinct Strategic Assessment Indigenous Impacts Report Volume 2: Traditional Owner Consent and Indigenous Community Consultation*. [Online] Available at: <http://www.dsd.wa.gov.au/documents/Appendix_E-2.pdf> (accessed 23 March 2011).

Koivurova T (2010) 'Limits and possibilities of the Arctic Council in a rapidly changing scene of Arctic governance', *Polar Record*, 46(2), 146–156.

Kotchen MJ and Young OR (2007). 'Meeting the challenges of the anthropocene: towards a science of coupled human-biophysical systems', *Global Environmental Change*, 17(2), 149–151.

Lachapelle PR, McCool SF, Patterson ME (2003) 'Barriers to effective natural resource planning in a "messy" world', *Society & Natural Resources*, 16(6), 473–490.

Leal PA (2007) 'Participation: the ascendancy of a buzzword in the neo-liberal era', *Development in Practice*, 17(4&5), 539–548.

Leemans R, Asrar G, Busalacchi A, Canadell J, Ingram J, Larigauderie A, Mooney H, Nobre C, Patwardhan A, Rice M, Schmidt F, Seitzinger S, Virji H, Vörösmart C, Young O (2009) 'Developing a common strategy for integrative global environmental change research and outreach: the Earth System Science Partnership (ESSP)', *Current Opinion in Environmental Sustainability*, 1(1), 4–13.

Muller S (2008) 'Indigenous Payment for Environmental Service (PES) opportunities in the Northern Territory: negotiating with customs', *Australian Geographer*, 39(2), 149–170.

Natcher DC, Davis S, Hickey CG (2005). 'Co-management: managing relationships, not resources', *Human Organization*, 64(3), 240–250.

Nelson A, Babon A, Berry M, Keath N (2008) 'Engagement, but for what kind of marriage?: community members and local planning authorities', *Community Development Journal*, 43(1), 37–51.

O'Faircheallaigh C (1999) 'Making social impact assessment count: a negotiation-based approach for Indigenous peoples', *Society & Natural Resources*, 12, 63–80.

O'Faircheallaigh C (2000) *Negotiating Major Project Agreements: The 'Cape York Model'*. AIATSIS Discussion Paper No 11 (Canberra: Australian Institute of Aboriginal and Torres Strait Islander Studies).

O'Faircheallaigh C (2002) *A New Approach to Policy Evaluation: Mining and Indigenous People* (Aldershot: Ashgate).

O'Faircheallaigh C (2004) 'Evaluating agreements between Indigenous peoples and resource developers'. In: Langton M, Tehhan M, Palmer L, Shain K (eds) *Honour Among Nations? Treaties and Agreements with Indigenous Peoples* (Melbourne: Melbourne University Press): 303–328.

O'Faircheallaigh C (2007) 'Environmental agreements, EIA follow-up and Aboriginal participation in environmental management: the Canadian experience', *Environmental Impact Assessment Review*, 27(4), 319–342.

O'Faircheallaigh C (2009) 'Effectiveness in social impact assessment: Aboriginal peoples and resource development in Australia', *Impact Assessment and Project Appraisal*, 27(2), 95–110.

O'Faircheallaigh C (2010) 'Public participation and environmental impact assessment: purposes, implications, and lessons for public policy making', *Environmental Impact Assessment Review*, 30(1), 19–27.

O'Faircheallaigh C and Corbett T (2005) 'Indigenous participation in environmental management of mining projects: the role of negotiated agreements', *Environmental Politics*, 14(5), 629–647.

Ostrom E and Cox M (2010) 'Moving beyond panaceas: a multi-tiered diagnostic approach for social-ecological analysis', *Environmental Conservation*, 37(4), 451–463.

Pacanowsky M (1995) 'Team tools for wicked problems', *Organizational Dynamics*, 23(3), 36–51.

Rodriguez-Izquierdo E, Gavin MC, Macedo-Bravo MO (2010) 'Barriers and triggers to community participation across different stages of conservation management', *Environmental Conservation*, 37(03), 239–249.

Schipper L and Pelling M (2006) 'Disaster risk, climate change and international development: scope for, and challenges to, integration', *Disasters* 30(1), 19–38.

Trebeck KA (2007) 'Tools for the disempowered? Indigenous leverage over mining companies', *Australian Journal of Political Science*, 42(4), 541–562.

Young OR, Berkhout F, Gallopin GC, Janssen MA, Ostrom E, van der Leeuw S (2006). 'The globalization of socio-ecological systems: an agenda for scientific research', *Global Environmental Change*, 16(3), 304–316.

14 Better learning

Alan Bond, University of East Anglia
Angus Morrison-Saunders, Murdoch University and North West University

Introduction

Learning lies at the heart of sustainability assessment. Gibson *et al.* (2005, p.187) note that: 'with the notion of sustainability itself, there is . . . no state to be reached', in other words, sustainability is a goal rather than an end state meaning it is always a case of being on a journey towards sustainability. To be aware of progress on that journey and to know or understand how to improve practices that will continue to make positive contributions to sustainability requires awareness and learning by all stakeholders in sustainability assessment.

Drawing on Chapters 4 and 5, where issues of temporal and spatial scales were considered, one of the key issues that emerges is that the learning involved (through the practice of sustainability assessment) must be renewed and revisited at several scales: spatially, from the individual and household scale, to neighbourhood and locality, institutional and national scales and across international settings and relationships; temporally, from the intergenerational to intragenerational, from decades to years, and from the future problems to the immediate issues. This requires conscious effort to recognise the validity and importance of scales which might otherwise be ignored.

It has been suggested that the existence of legal requirement for public participation within environmental assessment has led to 'state-sanctioned, deliberative spaces for civic interactions' (Sinclair *et al.*, 2008, p.415) which has learning potential. This is important for a number of reasons, but a key consideration in any *ex ante* assessment process is the level of uncertainty that remains in the predictions (de Jongh, 1988), which suggests not only that some level of experiential learning would be beneficial, but also some flexibility to adapt to unforeseen impacts (Sinclair *et al.*, 2008). Better engagement has been considered in Chapter 13 by O'Faircheallaigh and Howitt and this chapter will not seek to repeat what they have said, but it is clear that some elements of learning are dependent on the engagement process, and so when considering better engagement and better learning, one is not possible without the other. Practitioners, communities and institutional stakeholders are all able to be drawn into the learning that effective sustainability assessment nurtures.

Theory and types of learning

A range of sources present comprehensive consideration of types of learning relevant to assessment processes (see, for example, Sinclair and Diduck, 2001; Diduck and Mitchell, 2003; Nilsson, 2005; Chess and Johnson, 2006; Fitzpatrick, 2006; Sinclair *et al.*, 2008; Jha-Thakur *et al.*, 2009). However, it is important that some elements of the understanding in learning is synthesised here to provide a framework for considering better learning in sustainability assessment. Learning is generally considered to occur either on an organisational level or on an individual level (which is the learning experienced by the individual). Organisational learning is that which 'contributes to how that organization functions' (Fitzpatrick, 2006, p.159). Chess and Johnson (2006) suggest that organisational learning is essential to prevent historical inferences being lost, whilst Sánchez and Morrison-Saunders (2011) recognise that learning can be forgotten if not managed so that it is renewed, reinforced and transferred on. Some authors refer also to institutional learning, where 'institutions' are often defined as the rules of the game or habits that regulate interactions (Raina, 2003). However, for simplicity, and like Nykvist and Nilsson (2009), we regard organisations (the actors submitted to institutional rules) as being often inseparable and will consider institutional and organisational learning to be the same.

Most scholars researching learning within impact assessment adopt the theoretical framework of transformative learning, which is credited to Jack Mezirow (1981). This theory explains learning in adulthood and, specifically, the way in which it is different from learning in childhood (Diduck and Mitchell, 2003). It suggests that adults learn through transformations in frames of reference, and for learning in the context of sustainability, where the beliefs and values of the observer/learner are critical (to the understanding of what sustainable development actually means) this is a powerful argument for the relevance of the theory.

The theory of transformative learning is not universally accepted, and a particular criticism is that the transformation which needs to take place depends heavily on critical reflection, and this suggests a very rational process (as reflecting suggests rational thinking). This fits in very well with the original basis for Environmental Impact Assessment which was grounded on a 'better information' equals 'better decision' model dependent on rational decision making. However, a number of authors have discredited the reality of this (see, for example, Lawrence, 1997; Bartlett and Kurian, 1999; Cashmore *et al.*, 2004; Cashmore *et al.*, 2009), if not the underlying vision. Our position here is that for learning that can help to achieve more sustainable outcomes, accurate and relevant information is needed; the point we are making here is that rather than simply make a decision which is felt to be legitimised by the existence of evidence, decision makers need to learn from, and act on, that evidence.

Sinclair and Diduck (2001) distinguish three transformative concepts to describe how individuals learn:

1 Instrumental (or single-loop) learning which, as explained in Chapter 8, provides technical understanding and can facilitate changes in project designs with transforming the beliefs and values of the stakeholders.
2 Communicative learning which involves the derivation of meaning from social interactions. In a case study investigation in Canada, Sinclair and Diduck (2001, p.351) further subdivided this into '(i) insight into one's own interests, (ii) insight into the interests of others, (iii) communications strategies and methods, and (iv) social mobilisation'.
3 Emancipatory learning whereby resolution of disagreements takes place. Whilst there are a number of means whereby this can take place, Sinclair and Diduck (2001, p.341) refer to resolution through discourse which relies on: 'accurate and complete information; freedom from coercion; openness to alternative perspectives; ability to reflect critically upon presuppositions; equal opportunity to participate; and, ability to assess arguments as objectively as possible and to accept a rational consensus as valid'.

In Chapter 8, conceptual (or double-loop) learning was also introduced, drawing on the work of Nilsson (2005) in particular, as learning which leads to changes in beliefs (therefore fundamentally altering perspectives on policy and ways of achieving it). Conceptual learning relies on some element of transformation and so is consistent with the transformative theory of learning outlined above.

Berkes sets out three learning theories drawing on the work of Armitage *et al.* (2008) and applies these to the boundary between organisations and individuals within communities. He makes the point that individual and organisational learning are directly linked as 'participatory approaches seem to be central to learning by groups because they create the mechanism by which individual learning can be shared by other members of the group and reinforced' (Berkes, 2009, p.1697). The first of these learning theories, 'experiential learning', refers to situations where knowledge is created through experience or by 'learning-by-doing' (see, for example, Dougill *et al.*, 2006; Bond *et al.*, 2011). Secondly, 'transformative learning' is a reflective process that enables an individual to change their views, and can include either instrumental and conceptual learning, or both of them. Thirdly, social learning relies on engagement with others, typically in groups, to share experiences and debate with others, and to engage in a process of 'iterative reflection' (Berkes, 2009, p.1697).

In the context of both individual and organisational learning, Jha-Thakur *et al.* (2009) also refer to progressive learning to mean learning about sustainability assessment or learning through sustainability assessment (learning-by-doing). They combine this distinction with Bloom's taxonomy of learning levels (Bloom *et al.*, 1956) to categorise progressive learning as detailed in Table 14.1

Table 14.1 Progressive learning in sustainability assessment (adapted from Sinclair and Diduck, 2001; Jha-Thakur *et al.*, 2009)

Level of learning	Learning in sustainability assessment
Evaluation and Synthesis	Reflection and questioning personal, organisational or social beliefs and communication opportunities as a results of the sustainability assessment (learning through sustainability assessment)
Analysis and Application	Preparing or participating in a sustainability assessment, or the activity to which it applies (learning about sustainability assessment and learning through sustainability assessment)
Comprehension and Knowledge	Understanding about sustainability assessment, including legal requirements, engagement opportunities (learning about sustainability assessment)

Individual learning is distinct from social learning which is individual learning that is dependent on social interaction (Webler *et al.*, 1995). This is analogous to communicative learning as defined previously based on Sinclair and Diduck (2001). For social learning to take place, public participation needs to focus on achieving social learning and, Webler *et al.* (1995) argue, will lead to engagement which is widely perceived to have been successful.

It is clear that learning can take place in organisations or amongst individuals. Amongst individuals, social interaction is a critical vehicle for facilitating learning. It is also clear that there are different levels of learning and that higher levels of learning are more likely to lead to conceptual (or double-loop) learning which provides a better opportunity for realignment of individual or organisational norms to achieve a consensus regarding the goals of sustainability assessment. Evidence to date, however, suggests that whilst single-loop learning is common in assessment, double-loop learning is not (Fischer *et al.*, 2009). Some critical features of double-loop learning that have been highlighted are that:

1 it is not simply a case of understanding sustainability assessment; the process needs to be treated as a vehicle for learning
2 dialogue between individuals and groups is a necessary component of transformative change in views and beliefs
3 respect for different viewpoints and beliefs is essential for emancipatory learning which is necessary to reach agreement on sustainable development goals.

Examples of learning from sustainability assessment practice

Previous chapters have, even if unintentionally, highlighted examples of learning from sustainability assessment practice and other good examples are prevalent in the impact assessment literature more generally. Here we provide some illustrative examples with respect to different stakeholders in sustainability assessment processes: practitioners (including proponents, consultants and regulators), the public and researchers.

Learning in proponents

A good example of proponent learning from experience was provided in Chapters 7 and 10 in relation to the engagement of the Water Corporation in sustainability assessment. Within the context of a single assessment, for the South West Yarragadee Water Supply Development, the Water Corporation posed a 'rubbery proposal' as the starting point for the sustainability assessment. Through engagement with public stakeholders and experts, this proponent used an open active learning process to come up with the best design that would best meet sustainability objectives and arguably make the greatest contribution to sustainability for that project. Despite optimising that particular sustainability assessment, the government of Western Australia ultimately rejected the proposal, but that action provided an important learning opportunity for the Water Corporation who subsequently undertook a 50-year water-supply options sustainability assessment. This used a far more robust process that was ultimately well received by the Western Australian community at large. Thus, as long-term proponents, the Water Corporation have demonstrated a 'learning-by-doing' approach to sustainability assessment and provide an excellent role model in that regard to other proponents and practitioners. One could also argue that this represented double-loop learning in that the Water Corporation applied their new learning to completely different situations.

Learning in consultants

Proponents are typically assisted by the consultants they engage to carry out specialist sustainability assessment tasks on their behalf. While a proponent such as the Water Corporation may undertake numerous assessments each year (e.g. in Western Australia, it has carried out the most EIAs of any proponent operating in the state), many proponents may have 'one-off' involvement only in sustainability assessments (e.g. a company established to build a single land-development). For the latter, it is the specialist consultants that offer the expertise and continuity in experience and hence the opportunity to benefit from and to promote learning-by-doing. An example was provided in Chapter 10 of leadership by a consulting firm to encourage their clients to adopt a sustainability assessment approach that would otherwise be conventional EIAs as required by Western Australian legislation. This is an example of consultants as 'teachers' and something advocated by Weaver *et al.* (2008) for all sustainability assessment practitioners to advance practice.

Learning in regulators

A well-documented account of regulator learning with respect to EIA can be found in Sánchez and Morrison-Saunders (2011). In this instance the Environmental Protection Authority (EPA), through its combined responsibility for regulating and reporting on EIAs of all development projects assessed (including post-implementation monitoring and follow-up), environmental policy development, inputs to project licensing and pollution control, and responsibility for State of Environment reporting in Western Australia, represents a key vehicle for organisational learning over time. Sánchez and Morrison-Saunders (2011) provide examples of how their learning has been incorporated into ongoing modification and refinement of EIA processes in the state (and also identified a number of areas where knowledge was poorly managed, reducing the learning opportunities). Of particular note is the distinction identified between the knowledge managed by the EPA which is for internal use only, and that which is publically available, indicating that the opportunities for using internal knowledge to foster social learning are restricted.

The continuity of the EPA in Western Australian EIA practice has been core to its role as an educator and learner. In Canada, the creation of unique panels for each major assessment has the strength of tailor making the 'regulator' for each assessment but potentially reduces the organisational learning potential. Nevertheless in Chapter 11, Gibson argued that 'iterative learning' has taken place between different Canadian assessments over time (e.g. the Whites Point Quarry, Kemess North Copper-Gold Mine, Voisey's Bay Mine and Mill, and MacKenzie Gas Project respectively) which suggests that knowledge is passed on from one assessment to another and progressed. Iterative learning has not otherwise been classified by scholars, but we would suggest that it refers to a situation where progressive learning takes place between assessments rather than simply within them. Gibson (2011, p.243) also noted that while the contribution of the Joint Review Panel for the Mackenzie Gas Project assessment reported on in Chapter 11: 'is valuable as a working model for adaptation and advancement elsewhere, learning from many more applications will be important'. Openness and public accountability are core here so that not only successive panels can learn from the experiences and substance of their predecessors, but also the community. This Canadian approach, however, seems to favour single-loop rather than double-loop learning as there is no single organisation that remains in place to implement future learning – any opportunities for double-loop learning rely mainly on individuals maintaining roles across multiple panels or assessments over time.

Learning in the public

Public learning has long been identified as a fundamental element of assessment processes and outcomes. For example, in the context of EIA, O'Riordan wrote (in 1976):

> The nature of public response to EIAs and to participation in planning will depend upon innovations at all levels of learning. In its purest form, environmental education is simply 'consciousness raising' – the opening up of the mind to new levels of awareness and experience. This should encourage a greater sense of responsibility to and understanding of the interests of others, while improving personal knowledge of what is practically possible (and impossible). Properly speaking, an environmentally aware citizenry is a prerequisite of environmental impact assessment, since such people should be able to cope more effectively with the processes of social change and hence come to grips with the deepseated inconsistencies that infiltrate our values and question our actions.
>
> (O'Riordan, 1976, p.214)

This quote reinforces the point made earlier that there is a link between individual and social learning, whereby some level of individual knowledge is required in order to enter with confidence into a situation where social learning can take place. However, pointing to unambiguous empirical evidence of public learning through sustainability assessment is challenging. This is probably because the ability to demonstrate cause and effect in relation to learning arising from assessment activities in the absence of other societal events and influences on learning is virtually impossible to accomplish, and because until relatively recently, learning has very rarely been identified as a research issue in impact assessment. There is some limited evidence that learning has taken place in an English study that compared perceptions of citizens and consultants in two communities where EIA had been undertaken in the past (Robinson and Bond, 2003). The study was not investigating learning, but it is clear that some understanding of the process was present in citizens.

As previous chapters in the book have made clear, there is strong advocacy for public participation in assessment processes for sustainability and also much written in the literature to this effect. On the one hand Doelle and Sinclair (2006, p.204) state that: 'the more openly you engage the public, the earlier you do it, the better you do it, the better the project, and everyone wins', thereby implying that the collective learning process is actually essential for the accomplishment of sustainability outcomes. Similar sentiments are expressed in Sinclair *et al.* (2009) with respect to taking a community-based approach to strategic environmental assessment. Whilst community contributions to sustainability outcomes (i.e. harnessing the knowledge of the community) is one important element, Gibson *et al.* (2005, p.155) maintain that:

> sustainability assessment is not just about making better decisions. It is also about institutional and public learning. The sustainability-based decision criteria recognize the value of social-ecological civility and the need to deepen its roots in shared understanding and enriched capacity

for civic deliberation. Assessment processes, as much as the undertakings that results, should be designed to build this understanding and capacity.

A major conclusion of Gaudreau and Gibson (2010, p.242) with respect to their sustainability assessment of a biodiesel project in Barbados was that 'the assessment indicated that for the specific context at hand, the main benefit of biodiesel production is in promoting social learning rather than enhancing energy security and waste management'. This reflection identifies a role of sustainability assessment well beyond the specifics of an individual assessment.

Learning in researchers

Finally we suggest that researchers in the academic community play an important role in the learning associated with sustainability assessment – and are important targets for improved learning outcomes themselves (including many of our readers) and in fostering wider social learning. This book is one such example, with nearly all of its contributors being academics and seeking in their own way to understand and help others (i.e. proponents, consultants, regulators and community alike) to understand the intricacies of sustainability assessment theory, process and practice. Examples do exist of researchers learning from the applications of sustainability assessment in England (Thérivel *et al.*, 2009), from experience of policy instruments in South Africa (Rossouw and Wiseman, 2004), and from EIAs in Brazil (Bond *et al.*, 2010). This learning is passed on through academic literature. Other examples exist of researchers setting out to learn through involvement in a process of assessment (see, for example, Bond *et al.*, 2011), whereby greater experience is gained of social learning, which the researchers experience firsthand.

Better learning

The term 'adaptive co-management' is used by Berkes (2009) to describe an approach in which people work together to plan interventions, and actively reflect on the process both of the plan-making and plan-implementation, thereby learning from the experience. Adaptive co-management accommodates uncertainty and inaccuracy of prediction through adaptation, which is also critical in providing confidence to stakeholders that remedies will be put in place. Co-management is central to accommodating pluralism and must be responsive to 'ontological pluralism' (Suchet-Pearson and Howitt, 2006), which means acknowledging that different ways of knowing and learning exist in different cultural contexts and need to be accepted and included. Such an approach facilitates 'horizontal interaction among stakeholders, vertical interaction of communities with actors at different levels, and iterative learning' (Berkes, 2009, p.1698). From this approach can be extracted three key points to bear in mind when designing for better learning:

1 reflection is critical and needs to be included as a core part of sustainability assessment (relevant to all stakeholders in sustainability assessment)
2 engagement between stakeholders (including citizens) must occur in such a way that it enables constructive participation of all parties
3 sustainability assessment needs to facilitate emancipatory learning, involving the alignment of sustainability norms.

These three points tie in with the critical features of learning highlighted earlier on theory and types of learning and provide the basis for suggestions for better learning approaches in turn below.

Critical reflection in sustainability assessment

The need for critical reflection in sustainability assessment arises from the existence of uncertainty in *ex ante* decision-making tools which are necessarily generating uncertain predictions. The fact that such predictions may well turn out to be flawed, and that an assessment process was needed which did not end at the decision, but progressed through the life of a policy, plan or project, was recognised by Holling in the 1970s (Holling, 1978) in his suggestions for a process of 'Adaptive Environmental Assessment and Management'.

A number of authors have since recognised the inability of assessment processes in general to 'learn-by-doing', in that there is typically no critical reflection of previous practice. This has led to a lot of literature focussing on the need for follow-up in assessment (see, for example, Arts, 1998; Arts *et al.*, 2001; Morrison-Saunders and Arts, 2004; Marshall, 2005; Nilsson *et al.*, 2009), although it has not led to significant follow-up practice.

Sustainability assessment is achieving, in effect, normative goals. To be considered legitimate and successful, a diverse range of stakeholders need to feel that their views are reflected in the agreed norms. Assuming this is the case, critical reflection needs to focus on the sustainability process and more specifically on:

- the extent to which sustainability goals were agreed, including critical consideration of any reasons why this was not achievable for some stakeholders
- the extent to which the sustainability assessment process maintained the focus on the goals and delivered a project, plan or policy which achieved the agreed goals.

Critical and reflective engagement in sustainability assessment

The connection between better engagement and better learning is very clear, and the advice provided in Chapter 13 is critical to better learning. The fact that the outcome must still be open when the engagement takes place is a key underlying principle of any *ex ante* assessment process and is encompassed in numerous definitions of assessment. The International Association for Impact Assessment (IAIA), for example, define Environmental Impact Assessment as 'The process of identifying, predicting, evaluating and mitigating the biophysical, social, and other relevant effects of development proposals prior to major decisions being taken and commitments made' (International Association for Impact Assessment and Institute of Environmental Assessment, 1999).

Berkes (2009) argues that participatory approaches are necessary for learning in groups, and that social learning can proceed from instrumental learning, and extending through to conceptual learning in the process. Reference to Chapter 15 and the linkages identified between different categories of effectiveness reveals that a good sustainability assessment process needs to be designed such that sustainability imperatives are achieved. This book has indicated that the 'norms' for sustainable development should be those sustainability imperatives established by Bob Gibson in Chapter 1. However, these norms are not universally held and, whilst these may not be the same norms agreed as those selected as the goal of any particular sustainability assessment, some level of conceptual learning needs to take place in order to align the norms held by different stakeholders.

Overlapping with the need for reflection, social learning is a defining feature of adaptive management (Berkes, 2009), and can only be achieved through well-designed engagement. O'Faircheallaigh and Howitt use their experience of engagement with Indigenous peoples to argue in Chapter 13 how sustainability can be integrated transparently and fairly into decision making. Readers are referred back to that chapter for guidance on how to proceed. What is clear from the analysis in this chapter is that progressive learning will not take place without good engagement, and so a good sustainability assessment process is necessary to achieve sustainable outcomes.

Facilitating emancipatory learning

Sinclair *et al.* (2008) examined the use of educational techniques in environmental assessment in some Canadian jurisdictions, and they found that 'education was often an undervalued component of public participation' (Sinclair *et al.*, 2008, p.419). Research on public participation often finds that it is passive rather than properly engaging with members of the public and stakeholders. For example, Palerm (1999) used Habermas' theory of communicative action to develop principles for assessing the Aarhus Convention (United Nations Economic Commission for Europe, 1998), which has already been highlighted in Chapter 6 as a key driver for better engagement in assessment practice. Despite this, Palerm (1999, p.229) found that the Aarhus Convention fell short of the principles in four fundamental aspects: '(1) its need to ensure the participation of cognitively and lingu[istic]ally non-competent actors; (2) the need to have a two-way communication process;

(3) the need to ensure normative and subjective claims are adequately recognised; and (4) the need to establish conflict management procedures'. We can conclude from this that both practice and even the drivers of better practice fall short of the sort of engagement approaches highlighted in Chapter 13. Whilst O'Faircheallaigh and Howitt have indicated in their chapter how to improve engagement, they conclude that engagement 'is of limited benefit to communities unless it provides them with a substantial capacity to shape decisions that affect their well being, and can only enhance the quality of decision making if the sustainability of agreed outcomes over time is assured'. Similarly, Diduck and Mitchell (2003) highlight that as well as better engagement and earlier involvement, more open decision making is a crucial missing element that might lead to greater emancipation of stakeholders. They found that individual and organisational learning in relation to sustainable outcomes resulting from an assessment process tended to be undermined by the fact that the decision tended 'to confirm the ongoing dominance of conventional growth-oriented thinking' (Diduck and Mitchell, 2003, p.425). Thus, a key focus for better learning needs to be a process whereby decision makers are also learning through the process (rather than simply being advised of the results of the process) and have sufficient scope and authority to make decisions according to their learning. To an extent, this muddies the waters in terms of the vision of an objective decision maker making evidence-based decisions. The decision maker needs to be part of the assessment process to have enjoyed the opportunities for communicative learning.

How might decision makers be persuaded to be involved during an assessment rather than at the end? Quite simply, we believe that follow-up should become a mandatory (and enforced) part of sustainability (and other forms of) assessment; if they were, then they would have a vested interest in the process because there would be some expectation of action taking place to adapt to unforeseen circumstances. This makes a decision maker accountable in the long term.

Conclusions

At the outset of this chapter we set out how to demonstrate that sustainability assessment can contribute to better learning. Our starting point was that sustainability itself does not represent a fixed state and therefore a process of continuous learning and adaptation is inherent in its pursuit. Consequently any sustainability assessment process must intrinsically and explicitly contribute to learning for all stakeholders involved. As previous examples throughout this book have demonstrated, context matters and there is no single 'right' way to tackle sustainability assessment. As Gibson (2006, p.180) reminds us, 'the elaboration and implementation of sustainability assessment processes so far has involved a good deal of experimentation and learning-on-the-job'. We see this as a healthy way to approach sustainability assessment whether interest lies in theory, process or practical outcome dimensions. In doing so, a degree of humility will go a long way to advancing practice because, as Gibson *et al.* (2005, pp.89–91) espouse, a robust sustainability assessment process will facilitate 'learning from mistakes'; decisions and actions will not be perfect in the first instance.

More specifically though, components of the process such as public engagement and deliberation will naturally involve sharing of information and values amongst stakeholders and thereby automatically contribute to a learning process. The uncertainty inherent in any assessment process combined with the sustainability expectation to take a precautionary approach make it implicit that a sustainability assessment process should be designed to engender learning, and should accommodate adaptation over time as learning continues after the initial decision has been made. While learning should be seen as a fundamental objective or purpose of any sustainability assessment in this manner, it may not be something that can be enshrined easily in legislation and guiding processes; after all, the responsibility for learning to a large extent rests with the individual learner (be that citizens, organisations or society more generally). However, a firm legal requirement to engage in follow up is an essential way to establish sustainability assessment processes that will foster formal learning steps and opportunities.

References

Armitage D, Marschke M, Plummer R (2008) 'Adaptive co-management and the paradox of learning', *Global Environmental Change*, 18(1), 86–98.

Arts J (1998) *EIA Follow-Up. On the role of Ex Post Evaluation in Environmental Impact Assessment* (Groningen: GeoPress).

Arts J, Caldwell P, Morrison-Saunders A (2001) 'EIA follow-up: good practice and future directions – findings from a workshop at the IAIA 2000 conference', *Impact Assessment and Project Appraisal*, 19(3), 175–185.

Bartlett RV, and Kurian PA (1999), 'The theory of environmental impact assessment: implicit models of policy making', *Policy & Politics*, 27(4), 415–433.

Berkes F (2009) 'Evolution of co-management: role of knowledge generation, bridging organizations and social learning', *Journal of Environmental Management*, 90(5), 1692–1702.

Bloom B, Engelhart MD, Furst EJ, Hill WH, Krathwohl DR (1956) *Taxonomy of Educational Objectives: The Classification of Educational Goals. Handbook 1: Cognitive Domain* (London and New York: Longman Group Ltd).

Bond A, Dockerty T, Lovett A, Riche AB, Haughton AJ, Bohan DA, Sage RB, Shield IF, Finch JW, Turner MM, Karp A (2011) 'Learning how to deal with values, frames and governance in sustainability appraisal', *Regional Studies*, 45(8), 1157–1170.

Bond AJ, Viegas CV, Coelho de Souza Reinisch Coelho C, Selig PM (2010) 'Informal knowledge processes: the underpinning for sustainability outcomes in EIA?', *Journal of Cleaner Production*, 18(1), 6–13.

Cashmore M, Bond A, Sadler B (2009) 'Introduction: the effectiveness of impact assessment instruments', *Impact Assessment and Project Appraisal*, 27(2), 91–93.

Cashmore, M, Gwilliam R, Morgan R, Cobb D, Bond A (2004) 'The interminable issue of effectiveness: substantive purposes, outcomes and research challenges in the advancement of environmental impact assessment theory', *Impact Assessment and Project Appraisal*, 22(4), 295–310.

Chess C and Johnson BB (2006) 'Organizational learning about public participation: "Tiggers" and "Eeyores"', *Human Ecology Review*, 13(2), 182–192.

de Jongh, P (1988) 'Uncertainty in EIA', in P Wathern (ed.) *Environmental Impact Assessment: Theory and Practice* (London: Routledge), pp.62–84.

Diduck A and Mitchell B (2003) 'Learning, public involvement and environmental assessment: a Canadian case study', *Journal of Environmental Assessment Policy and Management*, 5(3), 339–364.

Doelle M and Sinclair AJ (2006) 'Time for a new approach to public participation in EA: promoting cooperation and consensus for sustainability', *Environmental Impact Assessment Review*, 26(2), 185–205.

Dougill AJ, Fraser EDG, Holden J, Hubacek K, Prell C, Reed MS, Stagl S, Stringer LC (2006) 'Learning from doing participatory rural research: lessons from the Peak District National Park', *Journal of Agricultural Economics*, 57(2), 259–275.

Fischer TB, Kidd S, Jha-Thakur U, Gazzola P, Peel P (2009) 'Learning through EC directive based SEA in spatial planning? Evidence from the Brunswick region in Germany', *Environmental Impact Assessment Review*, 29(6), 421–428.

Fitzpatrick P (2006) 'In it together: organizational learning through participation in environmental assessment', *Journal of Environmental Assessment Policy and Management*, 8(2), 157–182.

Gaudreau K and Gibson RB (2010) 'Illustrating integrated sustainability and resilience based assessments: a small-scale biodiesel project in Barbados', *Impact Assessment and Project Appraisal*, 28(3), 233–243.

Gibson RB (2006) 'Sustainability assessment: basic components of a practical approach', *Impact Assessment and Project Appraisal*, 24(3), 170–182.

Gibson RB (2011) 'Application of a contribution to sustainability test by the Joint Review Panel for the Canadian Mackenzie Gas Project', *Impact Assessment and Project Appraisal*, 29(3), 231–244.

Gibson RB, Hassan S, Holtz S, Tansey J, Whitelaw G (2005) *Sustainability Assessment: Criteria, Processes and Applications* (London: Earthscan).

Holling CS (ed.) (1978) *Adaptive Environmental Assessment and Management* (Chichester: John Wiley & Sons).

International Association for Impact Assessment and Institute of Environmental Assessment (1999) *Principles of Environmental Impact Assessment Best Practice.* [Online] Available at: <http://www.iaia.org/publicdocuments/specialpublications/Principles%20of%20IA_web.pdf> (accessed 25 November 2011).

Jha-Thakur U, Gazzola P, Peel D, Fischer TB, Kidd S (2009) 'Effectiveness of strategic environmental assessment – the significance of learning', *Impact Assessment and Project Appraisal*, 27(2), 133–144.

Lawrence DP (1997) 'The need for EIA theory-building', *Environmental Impact Assessment Review*, 17, 79–107.

Marshall R (2005) 'Environmental impact assessment follow-up and its benefits for industry', *Impact Assessment and Project Appraisal*, 23(3), 191–196.

Mezirow J (1981) 'A critical theory of adult learning and education', *Adult Education Quarterly*, 32(1), 3–24.

Morrison-Saunders A and Arts J (eds) (2004) *Assessing Impact: Handbook of EIA and SEA Follow-up*. (London: Earthscan).

Nilsson M (2005) 'Learning frames and environmental policy integration: the case of Swedish energy policy', *Environment and Planning C*, 23(2), 207–226.

Nilsson M, Wiklund H, Finnveden G, Jonsson DK, Lundberg K, Tyskeng S, Wallgren O (2009) 'Analytical framework and tool kit for SEA follow-up', *Environmental Impact Assessment Review*, 29(3), 186–199.

Nykvist B and Nilsson M (2009) 'Are impact assessment procedures actually promoting sustainable development? Institutional perspectives on barriers and opportunities found in the Swedish committee system', *Environmental Impact Assessment Review*, 29(1), 15–24.

O'Riordan T (1976) 'Beyond environmental impact assessment'. In O'Riordan, T and Hey R (eds), *Environmental Impact Assessment* (Hants, England: Saxon House) 202–221.

Palerm JR (1999) 'Public participation in environmental decision making: examining the Aarhus Convention', *Journal of Environmental Assessment Policy and Management*, 1(2), 229–244.

Raina RS (2003) 'Disciplines, institutions and organizations: impact assessments in context', *Agricultural Systems*, 78, 185–211.

Robinson M and Bond A (2003) 'Investigation of different stakeholder views of local resident involvement during environmental impact assessments in the UK', *Journal of Environmental Assessment Policy and Management*, 5(1), 45–82.

Rossouw N and Wiseman K (2004) 'Learning from the implementation of environmental public policy instruments after the first ten years of democracy in South Africa', *Impact Assessment and Project Appraisal*, 22(2), 131–140.

Sánchez LE and Morrison-Saunders A (2011) 'Learning about knowledge management for improving environmental impact assessment in a government agency: the Western Australian experience', *Journal of Environmental Management*, 92, 2260–2271.

Sinclair AJ and Diduck AP (2001) 'Public involvement in EA in Canada: a transformative learning perspective', *Environmental Impact Assessment Review*, 21(2), 113–136.

Sinclair AJ, Diduck A, Fitzpatrick P (2008) 'Conceptualizing learning for sustainability through environmental assessment: critical reflections on 15 years of research', *Environmental Impact Assessment Review*, 28(7), 415–428.

Sinclair AJ, Sims L, and Spaling H (2009) 'Community-based approaches to strategic environmental assessment: lessons from Costa Rica', *Environmental Impact Assessment Review*, 29(3), 147–156.

Suchet-Pearson S and Howitt R (2006) 'On teaching and learning resource and environmental management: reframing capacity building in multicultural settings', *Australian Geographer*, 37(1), 117–128.

Thérivel R, Christian G, Craig C, Grinham R, Mackins D, Smith J, Sneller T, Turner R, Walker D, Yamane M (2009) 'Sustainability-focused impact assessment: English experiences', *Impact Assessment and Project Appraisal*, 27(2), 155–168.

United Nations Economic Commission for Europe (1998) *Convention on Access to Information, Public Participation in Decision Making and Access to Justice in Environmental Matters* (Geneva: United Nations Economic Commission for Europe, Committee on Environmental Policy).

Weaver A, Pope J, Morrison-Saunders A, Lochner P (2008) 'Contributing to sustainability as an environmental impact assessment practitioner', *Impact Assessment and Project Appraisal*, 26(2), 91–98.

Webler T, Kastenholz H, Renn O (1995) 'Public participation in impact assessment: a social learning perspective', *Environmental Impact Assessment Review*, 15, 443–463.

15 Designing an effective sustainability assessment process

Alan Bond, University of East Anglia
Angus Morrison-Saunders, Murdoch University and North West University
Gernot Stoeglehner, University of Natural Resources and Life Sciences Vienna

Introduction

At this stage of the book it should be clear that sustainability assessment is very complex and sustainability assessment needs careful design if it is to help to achieve sustainable development. Chapters 3 and 8 have set the scene for considering what matters in sustainability assessment, while chapters 9–12 provided examples of some existing practice which is summarised in this chapter in order to highlight the critical areas which need to be addressed if practice is to be considered effective (judged by our own evaluation framework). Chapters 13 and 14 dealt specifically with the issues of *pluralism* and *knowledge and learning* and we recognise that these are critical to effective sustainability assessment, and have provided some insights on the best ways forward.

This chapter aims to help future practitioners navigate through the sustainability assessment design process. We argue that it is not (necessarily) appropriate to pick an off-the-shelf process, but that it is necessary to gain an understanding of the ways in which sustainability assessment will influence outcomes, values and perceptions so that it is designed to be fit-for-purpose. Indeed, the practice chapters have made it clear that in some countries, whilst the approaches taken have a sustainability remit, this in no way relies on formal or legal process requirements. In designing sustainability assessment, our argument is that an effective assessment process seeks to achieve the six imperatives of sustainability (as set out in Chapter 1 by Gibson), which must always be considered as criteria against which the process will be tested, through achieving effectiveness in all aspects of the evaluation framework. If efforts are not made to achieve effectiveness in all aspects, there will be a gap between the aspiration of the assessment process and the goals which are achieved.

Part 4 of this book (including Chapters 13–17) is about providing solutions to common problems which can arise during practice or be levelled at sustainability assessment by practitioners reluctant to change their ways and embrace the challenges such an approach demands. It is always easier to criticise than it is to find answers, but our aim here is to build on what we believe is already commendable practice and to take sustainability assessment to a new and higher level.

We begin by summarising what has been established, that is, what has been learned about process in practice through Chapters 7 and 9–12 in this book. Then, drawing on the literature, we establish principles for ensuring that sustainability assessment will be effective. In order to do this, there is a brief consideration of the theoretical framing of 'effectiveness' to try and ensure that the principles are not valid through one theoretical 'lens' only. We then move on to a conceptualisation of the linkages between the effectiveness criteria, and the sustainability imperatives established by Gibson in Chapter 1 – this conceptualisation allows further principles to be derived which accommodate all of the inter-linkages identified, and so provides a robust means of developing practice in sustainability assessment.

Summary of practice

Procedural effectiveness

In general, it seems that legal and administrative provisions for sustainability assessment procedures are well respected in the jurisdictions addressed in this book. If there is a particular requirement to do something, it can be expected that it will be done and there is little to suggest that current regulations are inadequate to the task of enabling a robust sustainability assessment process to be established within their remit. There is, however, some concern over the extent to which some tasks are conducted adequately (for example, in England there is a question mark over the extent to which alternatives are properly covered, and in South Africa there is some frustration over the actual sustainability benefits accruing from following set procedures), which indicates a direct relationship with substantive effectiveness. In short, sustainability is not likely to be achieved if appropriate procedural steps are not followed, but the evidence suggests that the procedural steps are generally being followed. Furthermore the ability to implement sustainability assessment in large part in the absence of formal expectations or requirements to do so, as the examples from Western Australia demonstrate, offer hope that practitioners can transcend procedural limitations and still achieve good outcomes. Overall what the situation demonstrates is that the procedural steps themselves are inadequate on their own for achieving sustainable outcomes at present.

Substantive effectiveness

The views on substantive effectiveness between jurisdictions vary. In Canada, England and Western Australia sustainability assessment has been found to change plans or even decisions in some cases (e.g. in Canada). In South Africa, sustainability assessment to date has not been seen to have directly changed plans, but to have indirectly had influence through raising knowledge and learning, and even influencing policy. So a direct link between the procedure undertaken and knowledge and learning can be seen in South Africa. In the other countries, despite the apparent positive influences of sustainability assessment, some frustrations remain that the changes were often minor, and the outcome was a move in the direction of sustainability, not actual attainment of 'sustainable development'. The challenge remains for all current and future practitioners of sustainability assessment to push sustainability thinking into mainstream and 'business-as-usual' approaches to impact assessment and decision making.

Transactive effectiveness

With respect to transactive effectiveness, in jurisdictions where forms of sustainability assessment are conducted as a legal requirement, the analysis suggests that proponents and governments express some discomfort over the time the assessments take and the expense. However there is also a feeling that improvements are possible and, indeed, likely as experience with sustainability assessment is gained and opportunities taken for tiering, whereby information and knowledge gathered at one level of decision making is passed down to the next, e.g. from plans down to projects (see, for example, Sánchez and Silva-Sánchez, 2008). In Western Australia, sustainability assessments are not mandatory and, where conducted, are considered by the proponents investing in them to be an efficient use of time and money. One could suggest that the difference between these two positions reflects one situation where sustainability assessments are undertaken for which (at least some) key stakeholders consider it valuable or necessary (in Western Australia), and another where sustainability assessment is undertaken much more widely because of legal requirements. In the latter case, it might be argued that no dialogue has taken place to debate the merits (or otherwise) of undertaking sustainability assessment (beyond the initial enactment of legal requirements for sustainability assessment), and the approach taken is less flexible than in Western Australia.

Normative effectiveness

Achieving normative effectiveness appears to be a major challenge in all jurisdictions examined in this book. The South African case is perhaps representative of practice in that the normative principles outlined by Gibson in Chapter 1 are all mandated as goals of the process. However, there is little, if any, measurement of the extent to which these goals are actually achieved. As already discussed at the outset of this chapter, measurement of success in conducting assessment is still largely focussed on procedural compliance, and not on the achievement of normative goals. In Canada, where there are many separate jurisdictions, practice varies and some (limited) success is reported. In both Western Australia and England, there are concerns about whether the process is really reversing unsustainable trends (evidence suggests it just slows the trend towards unsustainability), over trade-offs whereby the processes still seem to set economics against the environment, and over demonstrating mutually reinforcing gains. All jurisdictions claim openness and transparency with respect to the sustainability assessment process, but it is not clear that this openness is mirrored in the decision-making process in which firm choices for the trade-offs must necessarily be determined. The normative concept of sustainability on which a sustainability assessment is based influences the design of process as well as specific tools and methods applied, as described in Chapter 16 (where strong sustainability is pursued). Therefore, a certain amount of normative framing for designing process, tools and methods has already taken place, and effectiveness is then more likely when viewed through the same frame only. Svarstad *et al.* (2008) make exactly this point in relation to the application of the Drivers–Pressures–State–Impacts–Responses (DPSIR) framework, an approach for providing and communicating knowledge on the state of, and causal factors for, environmental issues that has been adopted by the scientific community and environmental agencies in most developed countries, the United Nations, Organisation for Economic Co-operation and Development, European Environment Agency, etc. (see, for example, European Environment Agency, 2006; Organisation for Economic Co-operation and Development, 2008). Therefore, judging normative effectiveness has at least two levels: on the level of the individual process it has to be assessed whether the norms adopted in the sustainability assessment agree with the norms laid out in the respective framework (for instance, the weighting of objectives, the consideration of trade-offs etc.). On the second level, there needs to be reflection on whether the sustainability assessment frameworks actually pursue sustainable development. The result of this reflection can be twofold and connects to issues of knowledge and learning: if frameworks are sufficient, but practice is not sufficiently promoting sustainability, education of practitioners to facilitate instrumental learning may resolve existing problems. If the normative base laid out in the sustainability assessment framework is not adequate and/or precise enough, the values and norms underlying the sustainability assessment have to be changed, which calls for conceptual learning. Drafting such sustainability assessment frameworks means making normative decisions which reduce the possibilities to accommodate normative pluralism in sustainability assessment, as it might not be possible to, for instance, pursue strong and weak sustainability at the same time, so these decisions have to be made cautiously and consciously.

Pluralism

The typical position in relation to the plurality of sustainability assessment is that certain authorities with a statutory remit to be consulted are engaged properly. However, the position regarding members of the public in England and Western Australia is that opportunities for comment are provided, but this does not necessarily equate to a genuine engagement and incorporation of views. Already, this raises concern that the divergent views of sustainability, and whether weak or strong sustainability framings should be pursued, are not being debated. In South Africa, there is no specific research to draw on, but it seems clear that those who do not like the decision are not happy about their engagement opportunities. In Canada, very large and contentious projects seem to cater well for stakeholder engagement, including participative hearings and funding to support citizen groups to gather evidence (this is not a cheap undertaking); smaller projects tend to suffer from the same problems as found in England and Western Australia. Clearly there is considerable scope to improve the plurality of sustainability assessment practice through greater engagement, although the examples provided in Chapter 13 of the complexity of engagement in culturally diverse pluralist settings highlight the incumbent challenges which lie ahead.

Knowledge and learning

All experience suggests that knowledge and learning has been enhanced by sustainability assessment. This learning may benefit regulators, proponents and other affected stakeholders alike; the burgeoning published literature on sustainability assessment (such as that cited in this book) points to learning and knowledge sharing in the academic and practitioner communities. However, there is no clarity over the exact mechanisms taking place, or the extent to which learning is instrumental or conceptual. Also of concern is that practice suggests that learning is largely restricted to the sustainability assessment community, including proponents, consultants, decision makers and consultees. Indeed, in Western Australia, the restriction of learning to this community was specifically noted, whilst in South Africa, the need to engage with 'value-based' and 'experiential' knowledge as opposed to traditional 'scientific' knowledge was recognised. Only in Canada has some evidence of participant learning been identified (with the inference being that the public are participants in the process). The goal for learning is clearly to achieve instrumental *and* conceptual learning in any sustainability assessment process, particularly within key stakeholders like decision makers, planners, and proponents on the one hand, and the interested and affected public on the other hand.

However, at present, there has not been research examining the extent to which sustainability assessment has led to institutional learning, and it will be a particular challenge for researchers to distinguish between true conceptual learning and political learning where improvements are purely symbolic.

Some reflections on the theory of effectiveness

All impact assessments have been founded on the principle that they provide evidence to decision makers so that they can make a better decision (although the notion of a 'better decision' is ambiguous given the plurality of views associated with the definition of sustainable development – as explained in Chapter 3). This is the rational underpinning for impact assessment which was explained by Cashmore and Kørnøv in Chapter 2. However, there are two key points that warrant initial consideration when thinking about sustainability assessment effectiveness. First, Cashmore and Kørnøv made it clear that there are many other models for the way that decision making works and that rational consideration of evidence by decision makers alone is not generally accepted to reflect reality (see, for example, Lawrence, 1997; Bartlett and Kurian, 1999; Cashmore *et al.*, 2009). Tools like sustainability assessment can be manipulated by decision makers (who may have individually compelling political pressures or agendas) which can undermine the principle that sustainability assessment is seeking to make better decisions. This point is dealt with in the following chapter which specifically addresses issues surrounding the integration of sustainability assessment into decision making. Second, decision makers in most (democratic) countries are elected, which is a governance principle which is rigorously defended. More pluralism (a principle which is being strongly advocated in this book) has been argued to lead to poor governance in situations where an emphasis on participatory and qualitative approaches presents poor evidence about sustainability impacts, based on a lack of technical understanding of cause and effect relationships (Kidd and Fischer, 2007; Bond *et al.*, 2011). What this means is that science and expertise is vital, but that this must be properly embedded in normative expectations. Sustainability assessment may contribute to make clear distinctions between the participatory negotiations of values, and the science-based examination of facts as well as aggregation of values and facts (rational-collaborative planning paradigm, see Chapter 16; Stoeglehner, 2010). Therefore, sustainability assessment might cater for a clear division of tasks between decision makers, involved stakeholders and public, science and expertise concerning issues of content as well as process design. In this way sustainability assessment might help to not only broaden the information base and to clarify the value base for the decision, but also to reveal power relations as well as inherent interests and hidden values, and increase normative and substantive effectiveness while facilitating learning and incorporating pluralism.

We accept there is validity in different theoretical perspectives, and agree with Cashmore *et al.* (2004) that there is a need for theory building to better understand the potential for impact assessment to contribute to sustainable development. Such theory building can usefully support practice 'on-the-ground' and help resolve issues and dilemmas faced by sustainability assessment practitioners.

Connecting criteria and imperatives

In order to be effective, the evaluation framework presented in Chapter 8 would suggest that the normative imperatives need to be achieved, and so sustainability assessment needs to be designed with this in mind. Figure 15.1 connects the imperatives with the evaluation framework criteria from Chapter 8. This conceptualisation of the cross linkages presents an opportunity to consider the key aspects of sustainability assessment that need to be considered in order to deliver a robust and effective process. As with any conceptualisation, it simplifies the actual situation in terms of linkages in order to reduce complexity and allow its use as something of a design tool. Nevertheless, it draws on our current understanding of practice, and the evaluation framework introduced in Chapter 8, in order to map out how the sustainability imperatives identified in Chapter 1 can be achieved. Figure 15.1 is not mapping out an integrated approach to sustainable development; Hacking and Guthrie (2008) commented on the complexity of the term 'integration', indicating that a link between procedure and 'integrating factors affecting sustainability' is questionable. However, in this context, the imperative is that there is an understanding of the importance of the interactions between the various influencing factors on sustainability, and that institutions typically struggle to work across disciplinary boundaries and institutional (or institutionalised) boundaries.

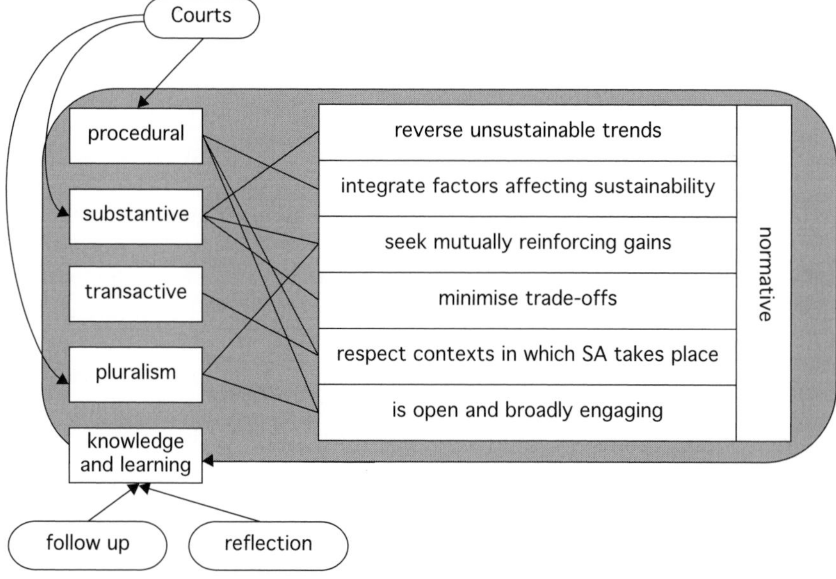

Figure 15.1

Inter-linkages between effectiveness criteria for sustainability assessment

Having presented the sustainability imperatives as a component of normative effectiveness, in line with the evaluation framework presented in Chapter 8, consideration has then been given to the inter-linkages with other effectiveness criteria. Procedural effectiveness has to be interpreted within the context of the sustainability assessment conducted in any particular jurisdiction. In Chapters 9–12 it was made clear that assessment can have a legal basis, or be conducted on a voluntary basis. Whichever of these is the case, clearly the particular steps or activities undertaken as part of the sustainability assessment have implications for the outcomes. Procedure does not change mindsets, but can demand dialogue to enhance interactions, and can impose methodological approaches which try to deal with interaction (as in the English case described in Chapter 9). Most important for procedure is that it not only supports individual rights of access to information and opportunity for consultation, but that it allows for learning experiences in dialogues between decision makers, planners (and proponents), stakeholders and the public, which is the subject of the next chapter. In democratic societies, the fact that environmental decision making is subject to a certain amount of public participation (normally information and consultation) is understood and, to a degree, enforceable by the courts. Both sustainability assessment procedures and practice, and the associated decision-making context can demand particular levels of involvement going beyond consultation – and there is a need to ensure these are truly open and broadly engaging.

Substantive effectiveness relates more to the outcomes of conducting sustainability assessment (see Chapter 8). This is an area where, traditionally, the courts do not intervene, both because it is rarely possible to link cause and effect in relation to outcomes with so many contributing variables, but also because decision makers are allowed discretion to make their decisions provided that procedures have been followed. The right to make those decisions has been mandated through the democratic process and it is not for the courts to intervene. In Chapter 6, it was demonstrated that the courts have little power to ensure sustainable decisions are made. For substantive effectiveness, the immediate goal has to be to move away from the existing trend of unsustainable development. To an extent, this can be exacerbated by sustainability assessment itself, where often it is predicated on the basis of ranking, whereby the least worst plan wins (see, for example, Thérivel *et al.*, 2009). In these circumstances, it is essential that the conduct of sustainability assessment be changed to a situation where the outcome is paramount and seen as the determining factor in decision making.

An issue here is related to these imperatives being normative and, therefore, the definition of what a sustainable outcome is may vary between those affected. Hence another critical determinant of substantive effectiveness is the extent to which the gains are mutually reinforcing. This overlaps with the imperative of minimising trade-offs in that more universal agreement that sustainable outcomes will be achieved depends on an attitude that all the pillars of sustainability are seen to be improving, rather than one or two at the expense of the others. This means not treating the sustainability assessment process as an end in itself, but treating the desired sustainability outcome as the end point which should influence assessment activity. We suggest that sustainability assessment practitioners must be ever vigilant on realising this and responding accordingly throughout any assessment activity; this involves a degree of reflection, as recommended by Burgess *et al.* (2007) and Chilvers (2007), which is dependent on follow-up providing some evidence for the outcomes resulting from sustainability assessment. In the context of the courts, the Strategic Environmental Assessment Directive (European Parliament and the Council of the European Union, 2001) states, in its Article 10 on Monitoring 'Member States shall monitor the significant environmental effects of the implementation of plans and programmes in order, *inter alia*, to identify at an early stage unforeseen adverse effects, and to be able to undertake appropriate remedial action'. It remains to be seen how courts will interpret this particular Article; the Directive was adopted in 2001 and Member States had until 21 July 2004 to adopt it. It has since been applied to plans and programmes which, for the most part, are in force for 10 to 15 years or longer. The implications of this Article text are, therefore, yet to be tested, but the inference is clearly that unpredicted impacts have to be identified (i.e., there must be some follow-up) in order to empower decision makers to carry out some form of reflection to see what the difference is between the vision and the reality after the action has been implemented, and either resolve the unforeseen impacts or choose to accept them.

Transactive effectiveness is pragmatic. There is little to be gained by spending more time and money on an assessment than is warranted by the decision context. The cost and resource implications of conducting sustainability assessment should be considered as part of the assessment process – whereby the social, economic and environmental benefits and costs of the assessment itself are considered part of the project, plan, activity etc. being considered. In most cases, the assessment should have a negligible footprint overall and so should not be influencing sustainability outcomes. However, in situations where a sustainability assessment may be so time-consuming and costly that it risks affecting the ability of the decision process to accommodate it, then it must necessarily be redesigned. Yet, transactive effectiveness is relative. It largely depends on the perception of sustainability assessment as being something useful. This perception is dependent on the other effectiveness issues, like substantive, normative etc. and on the experiences different stakeholders have already had with it. If sustainability assessment supports decision making in some effectiveness dimensions this creates 'ownership' (Stoeglehner *et al.*, 2009) of it by respective stakeholder groups and changes the value base for judging transactive effectiveness (see also Chapter 16). With respect to SEA practice, Thérivel (2004) provides some SEA design examples tailored to different time and money resourcing constraints such as how to carry out an SEA using a one person day of resources compared to 10 person days and 100 person days, with the level of effort therefore being proportional to the scale of the proposed activity. The idea is not to compromise on quality or outcome, but simply to pragmatically 'cut your cloth according to your means'; we suggest sustainability assessment could be guided by similar thinking.

Pluralism has been a central theme of this book, as it is critically bound with normative interpretations of sustainability (Jansen, 2003; Bond *et al.*, 2011; Bond and Morrison-Saunders, 2011). Sneddon *et al.* (2006) argue that pluralism must be embraced where multiple interpretations of sustainable development exist and, in the context of sustainability assessment, this means multiple interpretations of the goals of the process. Partly, this is aligned with the imperative that the process is open and broadly engaging, as this is the basis of integrating pluralism into the process. It also partly aligns with the imperative of seeking mutually reinforcing gains as the pluralism of views about the desired gains can lead to conflict. Accommodating pluralism will not automatically lead to sustainable outcomes and Chilvers (2007) stresses the need for reflection as a means of evaluating the connection between process and outcomes. Yet, incorporating pluralism might support sustainability assessment to reach a more holistic view of the values to be considered in decision making. This calls for participatory dialogues to agree on joint visions for sustainable development to be applied as a normative base in the assessment process. It also helps individual stakeholder or interest groups to reflect their own values and adjust them in order to reach sustainable development. In recent years, there has been a move for the courts to exercise some jurisdiction over public participation (i.e. facilitating some level of engagement with a population likely to have very different views) through instruments like the Aarhus Convention (United Nations Economic Commission for Europe, 1998), although it is clear from Chapter 6 that there is some way to go before there is a consistent legal basis to ensure equal funding for all parties affected by a particular decision.

Knowledge and learning are not directly aligned with the normative imperatives in the conceptualisation presented in Figure 15.1. However, we consider that progress towards sustainable development will be a learning process, and that considerable reflection will be needed on the extent to which normative imperatives are being achieved through the sustainability assessment process, and how the process is helping or hindering their achievement. Bond *et al.* (2010) found that knowledge needs to be managed in an impact assessment process, and a learning approach needs to be adopted in order to accommodate plurality of views and move towards sustainable decision making. Runhaar *et al.* (2010) indicate that actors involved in impact assessment selectively interpret knowledge generated through the process as they are subject to a variety of discourses which act as filters that sieve the relevant from the irrelevant. The end result can be that dominant discourses take over and knowledge which is inconsistent with them can be ignored (although knowledge is no less valid simply because it does not support certain discourses). They recommend reflection to ensure that knowledge is used in decision making. Sinclair *et al.* (2009) trialled a community-based approach to strategic environmental assessment based around critical reflection exercises, and found that the approach offered considerable potential, not least because the critical reflection reduced the power differentials between workshop participants, which otherwise might have favoured dominant discourses (as suggested by Runhaar *et al.*, 2010). One means of reflection familiar to impact assessment practitioners is the use of follow up (Arts and Nooteboom, 1999; Morrison-Saunders and Arts, 2004). This is the practice of checking on the performance of the impact prediction after the action that is the subject of the assessment has become operational. Sánchez and Morrison-Saunders (2011) found that, in Western Australia, knowledge was not captured from follow-up activities, and new knowledge was not generated; the conclusion here is that knowledge also needs to be managed. Hunsberger *et al.* (2005) recommend community involvement with monitoring and follow-up as this can produce results which are more locally meaningful, and can help to accommodate pluralism. An inescapable conclusion is that sustainability assessment must be designed such that learning ('instrumental' and 'conceptual' as explained in Chapter 8) is facilitated and knowledge is impartially managed and effectively used.

Better process?

In this chapter, the learning from the previous chapters in terms of the effectiveness criteria has been summarised, and an attempt has been made to conceptualise the links between the effectiveness criteria and sustainability imperatives. One thing that is clear is that no single model for effectiveness adequately explains decision making, and so if the sustainability imperatives are to be achieved, some attention needs to be paid to each of the criteria. Based on the analysis, a set of principles can be proposed which provide the basis for conducting any sustainability assessment practice; these principles are derived based on the linkages detailed in Figure 15.1 and explained above.

Principle 1: pluralism must be accommodated throughout the sustainability assessment, including the initial definition of desirable sustainability outcomes, and then throughout as implications of decisions are analysed. There is an opportunity for such engagement to be embedded in statute to allow enforcement through the courts. However there is nothing stopping practitioners from encouraging pluralistic engagement irrespective of legal arrangements.

Principle 2: the focus for the assessment must be on the sustainability of the outcome, not just the completion of expected steps in the assessment process. This means not stopping an assessment at the stage where outcomes are known, but continuing an iterative process of design and assessment until the outcomes are sustainable.

Principle 3: trade-offs and pluralism must not mix. Innovative approaches might be needed to design for gains across all pillars of sustainability, but this should be a requirement rather than an exception. One simple starting point to encourage practice in this direction would be to revise the mitigation hierarchy; currently this advocates 'avoid', 'reduce', 'abate', 'repair', 'compensate', and 'enhance' as the essential steps to be conducted, in the order written (Mitchell, 1997; Tinker *et al.*, 2005). However, we would advocate placing 'enhance' on the top of the mitigation hierarchy – the present emphasis on 'avoid' and 'minimise' obviously are not enough to meet the sustainability imperatives identified in Chapter 1.

Principle 4: there must be a presumption that sustainability assessment process and practice can always be improved. Sustainability or sustainable development is a moving target, and sustainability systems are dynamic; there will be no steady state that can be definitively categorised as being 'sustainable' but rather the opportunity to be 'more sustainable' should be evident. Reflection, adaptability and ongoing learning are requisite for the continuous improvement this principle anticipates.

Principle 5: process facilitates outcomes – and design facilitates good process. The discussion about process versus outcomes has to be approached from a different angle: process has to be designed in a way that principles 1–4 can be embedded in any particular sustainability assessment (see also Chapter 16). Procedural design is a means to embed the complexity of the linkages between the dimensions of effectiveness within the decision-making and assessment process.

These principles are few and straightforward. However, reference to Chapters 7 and 9–12 makes it clear that they are not universally applied, and from this we can begin to explain some of the weaknesses in effectiveness that were identified in those chapters and summarised in this chapter. Assessment needs to move on from the fixation with procedure and emphasise instead the outcome, acknowledging that there is much to be learned about how desirable outcomes might be achieved in any particular context. In terms of enforcement, this can be especially challenging given the complexity of multiple intervening factors which can affect sustainability outcomes. However, there are still opportunities to enforce the principles through procedural requirements for appropriate engagement, for follow-up, and for demonstrating that the approach taken in a given situation has maximised the gains across all pillars of sustainability. Such an approach may entail added burdens in terms of engagement and open-ended timescales for assessment when follow-up is considered. In the context of transactive effectiveness, we would argue that sustainability assessments are proportional to the activity being assessed, as advocated by Thérivel (2004). Sustainability assessment with an outcome focus rather than a procedural focus is likely to be a much better way to achieve sustainable outcomes and, as a welcome by-product, to avoid future litigation.

References

Arts J and Nooteboom SG (1999) 'Environmental impact assessment monitoring and auditing'. In Petts J (ed.) *Handbook of Environmental Impact Assessment – Vol.1 Environmental Impact Assessment: Process, Methods and Potential* (Oxford: Blackwell Science), pp.229–251.

Bartlett RV and Kurian PA (1999) 'The theory of environmental impact assessment: implicit models of policy making', *Policy & Politics*, 27(4), 415–433.

Bond A, Dockerty T, Lovett A, Riche AB, Haughton AJ, Bohan DA, Sage RB, Shield IF, Finch JW, Turner MM, Karp A (2011), 'Learning how to deal with values, frames and governance in sustainability appraisal', *Regional Studies*, 45(8), 1157–1170.

Bond AJ and Morrison-Saunders A (2011) 'Re-evaluating sustainability assessment: aligning the vision and the practice', *Environmental Impact Assessment Review*, 31(1), 1–7.

Bond AJ, Viegas CV, Coelho de Souza Reinisch Coelho C, Selig PM (2010) 'Informal knowledge processes: the underpinning for sustainability outcomes in EIA?', *Journal of Cleaner Production*, 18(1), 6–13.

Burgess J, Stirling A, Clark J, Davies G, Eames M, Staley K, Williamson S (2007) 'Deliberative mapping: a novel analytic-deliberative methodology to support contested science-policy decisions', *Public Understanding of Science*, 16(3), 299–322.

Cashmore M, Bond A, Sadler B (2009) 'Introduction: the effectiveness of impact assessment instruments', *Impact Assessment and Project Appraisal*, 27(2), 91–93.

Cashmore M, Gwilliam R, Morgan R, Cobb D, Bond A (2004) 'The interminable issue of effectiveness: substantive purposes, outcomes and research challenges in the advancement of environmental impact assessment theory', *Impact Assessment and Project Appraisal*, 22(4), 295–310.

Chilvers J (2007) 'Towards analytic-deliberative forms of risk governance in the UK? Reflecting on learning in radioactive waste', *Journal of Risk Research*, 10(2), 197–222.

European Environment Agency (2006) *Integration of Environment into EU Agriculture Policy – The IRENA Indicator-based Assessment Report. Report No.2 2006* (Copenhagen: European Environment Agency).

European Parliament and the Council of the European Union (2001) 'Directive 2001/42/EC of the European Parliament and of the Council of 27 June 2001 on the assessment of the effects of certain plans and programmes on the environment', *Official Journal of the European Communities*, L197, 30–37.

Hacking T and Guthrie P (2008) 'A framework for clarifying the meaning of triple bottom-line, integrated, and sustainability assessment', *Environmental Impact Assessment Review*, 28(2–3), 73–89.

Hunsberger CA, Gibson RB, Wismer SK (2005) 'Citizen involvement in sustainability-centred environmental assessment follow-up', *Environmental Impact Assessment Review*, 25(6), 609–627.

Jansen L (2003) 'The challenge of sustainable development', *Journal of Cleaner Production*, 11(3), 231–245.

Kidd S and Fischer TB (2007) 'Towards sustainability: is integrated appraisal a step in the right direction?', *Environment and Planning C*, 25, 233–249.

Lawrence DP (1997) 'The need for EIA theory-building', *Environmental Impact Assessment Review*, 17, 79–107.

Mitchell J (1997) 'Mitigation in environmental assessment – furthering best practice', *Environmental Assessment*, 5(4), 28–29.
Morrison-Saunders A and Arts J (eds) (2004) *Assessing Impact: Handbook of EIA and SEA Follow-up* (London: Earthscan).
Organisation for Economic Co-operation and Development (2008) *Environmental Performance of Agriculture in OECD Countries since 1990* (Paris: OECD).
Runhaar H, Runhaar PR, Oegema T (2010) 'Food for thought: conditions for discourse reflection in the light of environmental assessment', *Environmental Impact Assessment Review*, 30(6), 339–346.
Sánchez LE and Morrison-Saunders A (2011) 'Learning about knowledge management for improving environmental impact assessment in a government agency: the Western Australian experience', *Journal of Environmental Management*, 92, 2260–2271.
Sánchez LE and Silva-Sánchez SS (2008) 'Tiering strategic environmental assessment and project environmental impact assessment in highway planning in São Paulo, Brazil', *Environmental Impact Assessment Review*, 28(7), 515–522.
Sinclair AJ, Sims L, Spaling H (2009) 'Community-based approaches to strategic environmental assessment: lessons from Costa Rica', *Environmental Impact Assessment Review*, 29(3), 147–156.
Sneddon C, Howarth RB, Norgaard RB (2006) 'Sustainable development in a post-Brundtland world', *Ecological Economics*, 57, 253–268.
Stoeglehner G (2010) 'Enhancing SEA effectiveness: lessons learnt from Austrian experiences in spatial planning', *Impact Assessment and Project Appraisal*, 28(3), 217–231.
Stoeglehner G, Brown AL, and Kørnøv L (2009) 'SEA and planning: "ownership" of SEA by the planners is the key to its effectiveness', *Impact Assessment and Project Appraisal*, 27(2), 111–120.
Svarstad H, Petersen LK, Rothman D, Siepel H, Wätzold F (2008) 'Discursive biases of the environmental research framework DPSIR', *Land Use Policy*, 25(1), 116–125.
Thérivel R (2004) *Strategic Environmental Assessment in Action* (London: Earthscan).
Thérivel R, Christian G, Craig C, Grinham R, Mackins D, Smith J, Sneller T, Turner R, Walker D, Yamane M (2009) 'Sustainability-focused impact assessment: English experiences', *Impact Assessment and Project Appraisal*, 27(2), 155–168.
Tinker L, Cobb D, Bond A, Cashmore M (2005) 'Impact mitigation in environmental impact assessment: paper promises or the basis of consent conditions?', *Impact Assessment and Project Appraisal*, 23(4), 265–280.
United Nations Economic Commission for Europe (1998) *Convention on Access to Information, Public Participation in Decision Making and Access to Justice in Environmental Matters* (Geneva: United Nations Economic Commission for Europe, Committee on Environmental Policy).

16 Integrating sustainability assessment into planning: benefits and challenges

Gernot Stoeglehner and Georg Neugebauer, Institute of Spatial Planning and Rural Development, Department of Spatial, Landscape and Infrastructure Sciences, University of Natural Resources and Life Sciences Vienna

Introduction

Why should sustainability assessment be applied to planning – especially when we consider that many planning schemes strive for sustainability anyhow? And why should sustainability assessment be integrated into planning? In this chapter we investigate how these questions can be answered in the light of the effectiveness criteria laid out by Bond, Morrison-Saunders and Howitt in Chapter 8. As sustainability assessment is a relatively young domain with evolving, but still little, practice, this chapter presents a vision for how the effectiveness criteria can be addressed in integrated planning and sustainability assessment processes. To create the vision we build upon experiences gained with strategic environmental assessment (SEA) where, analogously to sustainability assessment, an assessment scheme was introduced into planning processes. The framework for this paper is derived from Stoeglehner (2010a), where the aspect of SEA-integration into planning is discussed, and Stoeglehner (2010b) where a concept for strategic planning and assessment methods is drafted that we propose should be applied in sustainability assessment to achieve substantive effectiveness.

We begin by categorizing the context of three types of plans in relation to sustainable development that assign different tasks to respective sustainability assessments for these plans. For our survey it is not important if plans fulfill certain formal criteria – e.g. leading to legally binding results – as long as objectives for further developments are formulated and action plans drafted grounded on baseline studies during the planning process. We move on to discuss content integration of planning procedures and sustainability assessment contents, followed by process integration. We then debate some methodological challenges for sustainability assessment integration into planning processes and introduce two examples for methods to deal with these challenges. The final section provides conclusions.

Sustainability in planning

In order to describe the relationship between sustainability and planning in a generic way we refer to environmental planning literature (see, for example, Weiland & Wohlleber-Feller, 2007; Grühn, 2004; Stoeglehner, 2009) and distinguish three types of plans:

- comprehensive plans
- sectoral environmental plans
- sectoral socio-economic plans.

Comprehensive planning includes all aspects of sustainable development, meaning that the environmental, social and economic pillars are considered and weighted in the planning process. Considering participation, normally information and consultation are guaranteed in formal planning instruments like spatial planning (urban and regional planning). This is especially true for EU Member States where an SEA has to be carried out, where according to the respective EU Directive (European Parliament and the Council of the European Union, 2001) consultation of the interested and affected public before the adoption of the plan or programme is (with some exceptions) obligatory. A more collaborative way of planning might be applied depending on the preferences of the decision makers. Examples are, for instance, land use plans or spatial development strategies. Informal comprehensive planning also takes place, for example, in Local Agenda 21 (LA21).

Sectoral environmental planning focuses on environmental protection, for example, management plans for national parks. The other pillars of sustainability might be addressed, e.g. when housing, tourism and agriculture are still allowed in or around the protected areas.

Sectoral socio-economic plans promote economic and/or social objectives and might cause considerable conflicts with environmental protection like, for example, energy planning or transport planning. Sometimes environmental protection is included as a sub-goal, which can often be linked to the objective to mitigate and/or compensate for negative environmental effects. For the latter category of plans environmental assessments are regularly applied (Willis and Keller, 2007) so that information and consultation rights for the public are at least granted in SEA and Environmental Impact Assessment (EIA).

Considering sustainability in planning, comprehensive plans might provide for the highest potential to integrate all sustainability pillars. Sectoral environmental plans can at least cover the environmental pillar sometimes with a clear decision to restrict socio-economic development in favour of environmental protection. In particular, plans implementing (socio-) economic objectives have a high potential to conflict with a strong sustainability view of sustainable development as they might often cause an overshoot of environmental capacity limits so that environmental sustainability cannot be fulfilled. This is especially a problem in industrialized countries where environmental overshoot is already present. Therefore, in order to converge to sustainability any planning measure must lead to a reduction of environmental pressures of the socio-economic system (Stoeglehner, 2010a).

Integrating planning and sustainability assessment contents

The debate about integration of planning and sustainability assessment contents is highly controversial. Critical voices point out that sustainability assessment might lead to trade-offs of environmental concerns against social and economic benefits (see, for example, Dalkmann, 2005; Morrison-Saunders and Fischer, 2006; Stoeglehner, 2010a). Some evidence was gained from English experiences that this is actually happening in sustainability assessment practice (Thérivel *et al.*, 2009). From our point of view this integration of contents (with the risk of accepting environmental effects at the expense of socio-economic gains) will happen in any decision-making process, so that the sustainability assessment or not-sustainability assessment question is less important than the question of how the contents are integrated, which is mainly a matter of the value base of the planning and assessment process. Planning and decision making implies an approach to 'connect a level of values and a level of facts (Fürst and Scholles, 2001) influenced by power relations, concrete situations and actor constellations (Scharpf, 2000)' (Stoeglehner, 2010a, p.220f) which means that values, facts and aggregation rules have to be clearly laid out for sound decision making. Stoeglehner (2010a) promotes a rational-collaborative planning paradigm where values and aggregation rules are collectively negotiated, whereas the elaboration of facts and the aggregation itself are science based. Therefore, abstaining from sustainability assessment cannot prevent trading off certain sustainability issues for others. The arguments call for a clear and deliberate structure for the content integration of all pillars of sustainability in the planning and decision-making process with or without sustainability assessment.

As sustainable development is a normative concept laid out in the Rio Declaration (United Nations, 1992) with considerable room for interpretation we propose that the problem of 'trade-offs' (normative principles, see Chapter 8) between the three pillars depends on the sustainability concept underlying the decision-making process. Therefore, content integration of all pillars of sustainability can only be properly designed if the sustainability concept underlying the planning and sustainability assessment process is explicitly agreed beforehand, as the normative base for the planning and assessment process not only determines the results of the process, but also the methods and techniques chosen. Gibson (Chapter 1) as well as Bond and Morrison-Saunders (Chapter 3) reviewed different sustainability concepts and made the diversity of interpretations of sustainability visible. In order to draft our model of content integration we rely on the concept of 'strong sustainability' depicted as the 'egg of sustainability' (see Figure 12.1) with the yolk (further differentiated to society and economy) being the anthroposphere and the white of the egg being the environment (Birkmann, 2000). The 'egg of sustainability' clearly shows the interrelations between the three pillars: the environment represents the overall system. Society is a sub-system of the environment and economy a sub-system of society. If one sub-system overshoots the limits of the surrounding system(s) the overall system is destabilized and endangered. Sustainability can only be reached if the sub-systems fulfill sustainability criteria and stay within the limits of the superior systems (see e.g. Birkmann, 2000; Narodoslawsky and Stoeglehner, 2010). Therefore, this model offers a clear system and succession of considerations to avoid trade-offs.

The 'sustainability egg' can be operated along the 'indicator pyramid' (Figure 16.1) according to Stoeglehner and Narodoslawsky (2008): the indicator pyramid has to be read top down and symbolizes that the information load increases throughout the planning and decision-making process. It is proposed that on the level of a pre-assessment a screening of alternatives using general indicators takes place based on an 'un-sustainability judgment': for the environmental pillar such an indicator can be the ecological footprint (applying modified calculation methods) to assess if the alternatives comply with defined limits of environmental capacity. Only alternatives that pass this pre-assessment are assessed in more detail using specific indicators, addressing specific environmental, social and economic issues. The indicator pyramid also shows that in line with the sustainability egg the logic of weighting means that first the environmental assessment has to be passed, second the social assessment, third the economic assessment. In this way planning alternatives are chosen that are environmentally and socially acceptable as well as economically affordable. This pre-assessment might be used to define all alternatives that have the potential to be sufficiently sustainable, so that they can be processed in the detailed assessment. This concept also decreases the information load on decision makers, stakeholders and the public so that it supports focused participatory debate and sustainability oriented decision making.

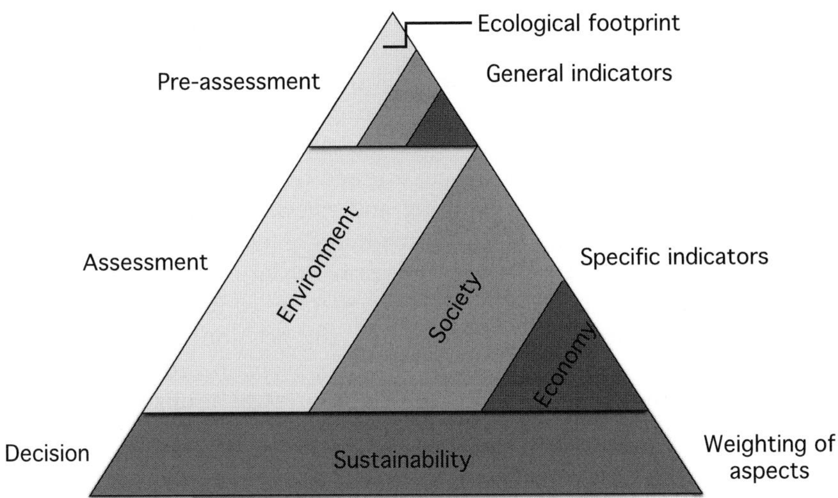

Figure 16.1

The 'indicator pyramid' (Stoeglehner and Narodoslawsky, 2008)

From this model of integrating environmental, social and economic issues into one decision geared towards sustainable development it can be seen that sustainability assessment for the different types of plans introduced above has different meanings, and, therefore, different contributions of sustainability assessment to planning processes can be expected. In comprehensive plans the consideration of all pillars of sustainable development should be guaranteed, yet it might remain unclear whether the weighting of issues corresponds to sustainable development, e.g. that all pillars are addressed in the planning report, but only a certain pillar is considered in the final decision. Comprehensive plans are specific as objectives and contents for all sustainability pillars should be addressed in the planning and assessment process. Clear and transparent rules should be applied which can be documented in an explanatory report of the planning process or in a sustainability assessment report. Therefore, a contribution of a sustainability assessment for comprehensive plans to more sustainable development might be associated with the following benefits (analogous to SEA contributions after Stoeglehner, 2010a): proof of sound methodological decision making concerning the integration of all pillars of sustainability; traceability and transparency of the process; clear definition of the facts, the values and the planning and assessment techniques applied; appraisal of cumulative impacts; proposal of mitigation and compensation measures.

Considering sectoral environmental plans or sectoral socio-economic plans, a sustainability assessment might provide the incentive to introduce all pillars of sustainability into the planning processes, which means socio-economic issues in environmental plans or environmental considerations in socio-economic plans. Sectoral plans differ from comprehensive plans as only a small spectrum of objectives is actively strived for, but all other objectives have to be addressed to improve the situation or mitigate or compensate for negative effects. As any sectoral planning process has effects on all pillars of sustainability, all issues have to be surveyed, assessed and weighted in a planning and assessment process. The planning and assessment results have to be integrated into one single decision about the adoption of the preferred planning alternative. A sustainability assessment can contribute to sectoral planning substantially by defining the affected issues within the other pillars, by assessing the effects and by supporting the decision in line with the indicator pyramid.

Sustainability assessment shaped in this way might lead to substantive effectiveness (Chapter 8) changing actions and outcomes as it clearly refers to normative principles (Chapter 8) that minimize or prohibit trade-offs. Whether prevailing unsustainable trends are reversed depends on the value base applied in the planning and sustainability assessment process and, *inter alia*, which values or thresholds are used for passing the pre-assessment.

Integrating planning and sustainability assessment processes

For the discussion of process integration we can refer to SEA: when considering generic planning process schemes and blending them with generic assessment process schemes it is obvious that compliance is high (see, for example, Fischer *et al.*, 2002; Partidário *et al.*, 2008; Stoeglehner, 2004). Evaluation of experiences with SEA reveals that process integration and timing are important preconditions to guarantee effective SEA (Retief, 2007; Runhaar and Driessen, 2007; Stoeglehner, 2010a; Thérivel and Minas, 2002): results of the assessment have to be readily available to influence the decision-making process, which means *before* the decision is made. Only a few voices in the SEA debate call for separated planning and assessment approaches (see, for example, Elling, 2009). In our understanding, planning and assessment should be merged into one process in line with the majority of SEA scholars which can be also argued from the perspective of communicative and collaborative planning theory as well as learning theory.

According to communicative and collaborative planning theory (see, for example, Healey, 1992; Mueller, 2004) planning decisions should strive for consensus building in communication processes between citizens, planners and decision makers. Visioning is an important part of the planning process. As assessment processes can be designed as a communicative process (as stated by Richardson, 2005 for SEAs) and integrating values and objectives into the planning process might become the most important task of an assessment, even more important than the write-up of an assessment report (as is the case for SEA according to Brown and Thérivel, 2000; Partidário, 1996; Stoeglehner, 2010a). Taking up the discussion of 'process versus outcomes' (Chapter 3) we do not propose that process is more important than outcomes, but that process has to be designed in order to achieve sustainable outcomes.

We approach this issue from a learning theory perspective: a planning and assessment process can be understood as a collective learning process within groups of relevant decision makers, stakeholders and planners (Stoeglehner, 2010a) that allows for learning about facts and learning about values. By reflecting on the consequences of recent developments and proposed actions two ways of learning can be induced – single-loop (or instrumental) learning and double-loop (or conceptual) learning (Argyris, 1993, cited by Innes and Booher, 2000; Stoeglehner, 2010a). As instrumental learning reflects the proposed actions and their consequences with a given value base it may lead to adoptions of measures without questioning the vision and the underlying values. Therefore, improvements of measures as well as provision of compensation for negative effects are possible, but not changes of the vision. In conceptual learning, the values and vision are also questioned and undesired consequences might lead to a general change of the vision, rather than simply an adaptation of measures. Instrumental learning can be identified with assessment goals related to the mitigation of negative impacts, but the change of societal processes which is necessary to achieve sustainable development (Kanatschnig and Weber, 1998; Meadows *et al.*, 1992) can be supported by conceptual learning (Stoeglehner, 2010a). Therefore, we propose that a certain kind of process is necessary to achieve sustainable outcomes, and this is a process supporting conceptual learning. Furthermore, these processes need a framework for content integration as laid out above to achieve sustainable outcomes.

Summing these theory considerations up, participation is an important feature of sustainability assessment. Therefore, we suggest participation should be added as the fourth pillar of sustainable development addressing the planning and sustainability assessment process. The Rio Declaration refers to participation in different sections, e.g. for Local and Regional Agenda 21 (LA21; United Nations, 1992, Chapter 28). According to the Austrian Local Agenda 21 guidelines (ExpertInnengruppe LA21, 2010) five levels of participation can be distinguished with different intensities of interaction spanning from information to self-responsible implementation of visions and action plans by the participating public (see Figure 16.3). From these five levels the first three have to be fulfilled to call a process participatory according to the LA21 framework. Following these principles we argue that information and consultation procedures normally applied in planning and assessment processes are not sufficient to promote participation as the fourth pillar of sustainable development.

From these considerations it can be concluded that sustainability assessment should be already in operation when the value base is agreed and reflected in the planning process in such a way that participation as the fourth pillar of

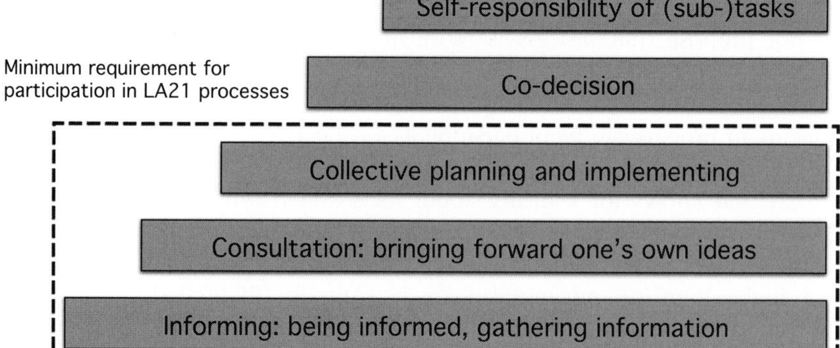

Figure 16.2

Levels of participation (source: ExpertInnengruppe LA21, 2010)

sustainable development can be guaranteed. In analogy to the SEA literature (Stoeglehner, 2004; Stoeglehner *et al.*, 2009; Stoeglehner, 2010a) process integration can be 'technically' proven by overlapping planning and assessment steps (see the first two columns of Figure 16.4), as shown on linear, planning process schemes grounded in rational-planning theory.

We doubt that such integrated planning and assessment schemes lead to the desired learning experiences if information and consultation are the only ways of interaction between planners and decision makers, and the public. Collective planning and implementation of visions and action plans according to Figure 16.3 should be the minimum standard of participation. Information and consultation – normally affiliated with formal planning and assessment instruments – is important for keeping individual and public rights and interests but will not lead to social learning as well as communicative and collaborative planning outlined above. In order to achieve instrumental and conceptual learning, participation has to be facilitated on the one hand, whilst planners and decision makers must be willing to undergo iterative planning loops between the steps of the planning and assessment process on the other hand.

Therefore, we argue that process qualities of sustainability assessment might become as important – but not more important – than content qualities defined above and depicted in Figure 16.4 (third column). These process qualities might endorse pluralism, knowledge and learning (Chapters 3, 7, 8) in a way that all interests, stakeholders, organizations and/or persons present in, and relevant for, the process have a chance to bring in their knowledge and values in the negotiation of a shared value base and the agreement of aggregation rules (see the section on the rational-collaborative planning paradigm). Depending on the results of the assessment, values and actions can be re-negotiated in

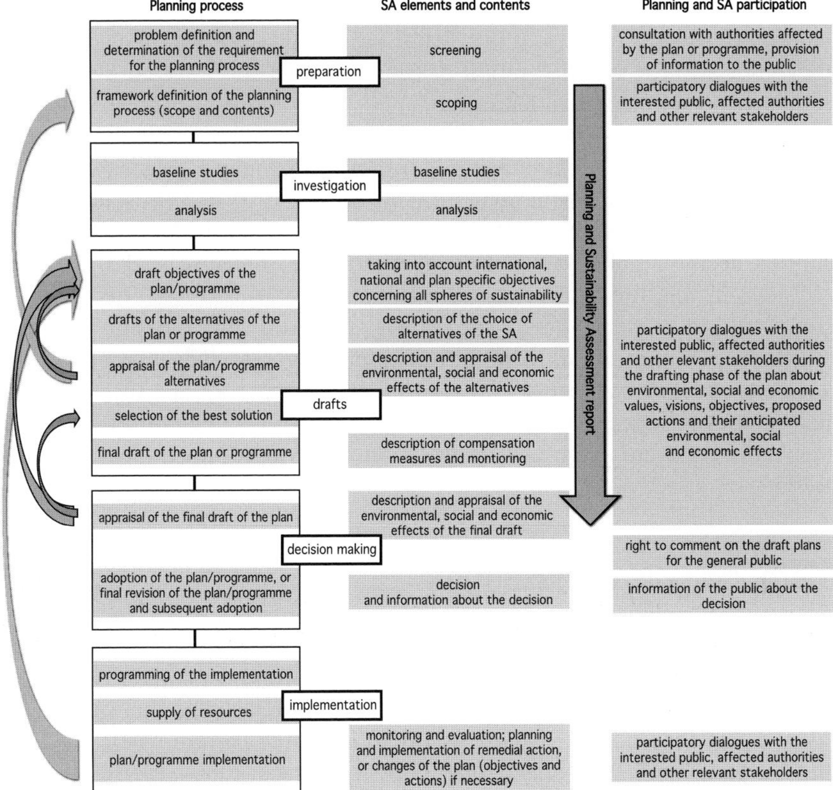

Figure 16.3

Integrated planning and sustainability assessment process with participatory dialogue (after integrated planning and SEA processes according to Stoeglehner, 2004; Stoeglehner *et al.*, 2009; Stoeglehner, 2010a)

conceptual learning processes. The only framework that has to be followed is the proposed hierarchy of environmental, social and economic values according to the indicator pyramid. Furthermore, Figure 16.4 might provide a framework for assessing procedural effectiveness (Chapter 8) as the necessary steps for the integrated, participatory planning and assessment process are defined.

In processes for comprehensive plans it is likely that many interests related to all pillars of sustainability are present. Within the sustainability assessment it should be guaranteed that all interests, also the missing ones, are advocated in collective planning (and implementation) and that broad participation takes place if this is not guaranteed in the planning process itself. In planning processes for sectoral plans (no matter if environmentally or socio-economically driven) it is more likely that only a narrow spectrum of stakeholders is involved in the drafting phase. Therefore, the sustainability assessment process might not only cater for content integration as discussed above but also for the integration of the interested and affected public as a prerequisite for conceptual learning.

Finally, we propose in line with Stoeglehner *et al.* (2009) that the concept of sustainability assessment integration into planning might create 'ownership' of the sustainability assessment by the planners, decision makers, stakeholders and the public involved – which means that sustainability assessment might improve decision making apparent to all groups and, therefore, lead to accepted decisions in the direction of overall sustainability. In this way the involved groups might not only own the values, methods and outcomes of the integrated planning and sustainability assessment process, they might also wish to carry out further sustainability assessments. In this way, the sustainability assessment concept proposed for content and process integration might also support what Bond, Morrison-Saunders and Howitt (Chapter 8) refer to as 'transactive effectiveness'.

Methodological challenges

Methodological challenges for sustainability assessment are manifold and are related to the methods and techniques applied for carrying out the assessment. Different approaches, techniques and methods for tackling the assessment tasks of sustainability assessment are discussed in the literature (e.g. Hacking and Guthrie, 2008; Pope *et al.*, 2005; Pope and Grace, 2006; Stoeglehner and Narodoslawsky, 2008; Narodoslawsky and Stoeglehner, 2010). If participation as the fourth pillar of sustainable development is introduced into the sustainability assessment concept, techniques of process design and moderation have to be added.

In order to overcome methodological challenges and to give incentives for the development of respective methods and tools we discuss two examples of sustainability-oriented decision making from Austria. The first example concerns informal comprehensive planning via Local Agenda 21. In Austria quality guidelines for LA21 are established that formulate 34 content criteria as well as process criteria. The content criteria cover all three pillars of sustainable development and are formulated as objectives. The underlying idea is that if a criterion is fulfilled, the local (or regional) community will automatically take direction towards sustainable development. In order to comply with the minimum standards at least 50 per cent of the criteria and at least one criterion of each pillar have to be covered. Furthermore, from the above quoted levels of participation, at least the levels of information, consultation, collective planning and implementation have to be guaranteed. The process is bottom-up, based on voluntary work and is supported by a professional process attendant. The municipalities can obtain subsidies for the process attendant and certain expenses for the process, the implementation of measures is not funded by the LA21 programme. Fulfilling the minimum requirements is obligatory to obtain subsidies for the LA21 process (ExpertInnengruppe LA21, 2010).

Although the framework is highly informal the outcomes are remarkable. Within the framework of the Central Europe project 'Vital Landscapes' (www.vital-landscapes.eu) the authors carried out an analysis of 21 Austrian LA21 vision statements. During an international Vital Landscapes expert workshop a set of criteria for sustainable landscape development was developed and then, *inter alia*, applied to the Austrian case studies. Table 16.1 reveals the results of the case study comparison in a matrix. The presence ('Y') or absence ('N') of criteria can be interpreted as factors for success or failure to achieve sustainable development.

More than half of the assessed strategies addressed all the aspects. From the seven criteria attached to the thematic area 'environment and nature' all case studies deal with natural capital/heritage and most of the case studies with regional resource cycles, renewable energy sources and climate change adaptation and/or mitigation. All the processes assessed touch upon issues of quality of life, social capital and cultural capital/heritage in the 'social issues & culture' theme, whereas consciousness/awareness of landscape is not subject to consideration in quite a few case studies. In the 'economic issues' theme (economic capital, regional resources, regional labour market and regional economic cycles/regional income generation), most of the assessed processes deal with four out of five criteria.

In a LA21 process some kind of assessment takes place at least in the choice of topics, analysis of the situation and choice of visions and actions. By defining certain rules about how many criteria of process and content of the LA21 must be achieved, such an assessment can:

- build a common understanding of what sustainability is and which kind of activities could contribute to sustainable development
- define a minimum standard for process design and contents of a certain LA21 process
- give ideas to the involved local stakeholders about what could be elaborated in a LA21 process
- check if a LA21 fulfills the criteria for getting co-financed by the authority responsible for LA21.

The method is simple, is confined to yes/no decisions and by filling in the assessment form the assessment can be completed easily. This procedure is in principle feasible for processes that are driven by volunteer work. The assessment procedure prompts the assumption that setting a certain amount of actions concerning environmental, social and economic topics is enough to achieve sustainable development. It also promotes a kind of triple-bottom-line approach as implementing measures addressing all pillars of sustainability already creates a local sustainability strategy. One can argue that these kinds

Table 16.1 Case study comparison (source: own processing), [key: NÖ = Lower Austria, OÖ = Upper Austria, Szbg = Salzburg, Stmk = Styria, Vbg = Vorarlberg]

		NÖ, Ardagger	NÖ, Harmannsdorf	NÖ, Prellenkirchen	OÖ, Altmünster, Neukirchen, Reindlmühl	OÖ, Bad Zell	OÖ, Freistadt	OÖ, Gutau	OÖ, Lasberg	OÖ, Molln and Leonstein	OÖ, Mühlviertler Alm	OÖ, Ottensheim	OÖ, Steinbach/Steyr	OÖ, St. Stefan am Walde	OÖ, Vorderstoder	Szbg, Bergheim	Szbg, Thomatal	Stmk, Almenland	Stmk, Kleinregion Gleisdorf	Stmk, Hohenbrugg-Weinberg	Vbg, Region AmKumma	Vbg, Langenegg
Environment and nature	A1 natural capital/heritage	Y	Y	Y	Y	Y	Y	Y	Y	Y	Y	Y	Y	Y	Y	Y	Y	Y	Y	Y	Y	Y
	A2 ecological carrying capacity	Y	Y	Y	Y	Y	N	N	Y	N	Y	Y	Y	Y	N	Y	Y	Y	Y	Y	Y	Y
	A3 landscape preservation	Y	Y	N	Y	Y	N	N	Y	Y	Y	N	Y	Y	Y	Y	Y	Y	Y	Y	Y	Y
	A4 landscape development	Y	Y	N	Y	Y	N	Y	Y	N	Y	N	Y	Y	Y	Y	Y	Y	Y	Y	Y	Y
	A5 regional resource cycles	Y	Y	Y	Y	Y	Y	Y	Y	Y	Y	Y	Y	Y	Y	N	Y	Y	Y	Y	Y	Y
	A6 renewable energy sources	Y	Y	Y	Y	Y	Y	Y	Y	Y	Y	Y	Y	Y	Y	Y	Y	Y	Y	Y	N	Y
	A7 climate change adaptation and/or mitigation	Y	Y	Y	Y	Y	Y	Y	Y	Y	Y	Y	Y	Y	Y	Y	Y	Y	Y	Y	N	Y
Social issues and culture	B1 quality of life	Y	Y	Y	Y	Y	Y	Y	Y	Y	Y	Y	Y	Y	Y	Y	Y	Y	Y	Y	Y	Y
	B2 social capital	Y	Y	Y	Y	Y	Y	Y	Y	Y	Y	Y	Y	Y	Y	Y	Y	Y	Y	Y	Y	Y
	B3 cultural capital/heritage	Y	Y	Y	Y	Y	Y	Y	Y	Y	Y	Y	Y	Y	Y	Y	Y	Y	Y	Y	Y	Y
	B4 demographic change	Y	Y	Y	Y	Y	Y	Y	Y	N	Y	Y	Y	Y	Y	Y	Y	N	Y	Y	Y	Y
	B5 consciousness/awareness of landscape	Y	N	N	Y	Y	N	N	Y	Y	Y	N	Y	Y	Y	Y	Y	Y	Y	Y	Y	Y
	B6 local/regional identity	Y	Y	Y	Y	Y	Y	Y	Y	Y	Y	Y	Y	Y	N	N	Y	Y	Y	Y	Y	Y
Economic issues	C1 economical capital	Y	Y	Y	Y	Y	Y	Y	Y	Y	Y	Y	Y	Y	Y	Y	Y	Y	Y	Y	Y	Y
	C2 multifunctional agriculture	Y	Y	Y	Y	Y	Y	Y	Y	Y	Y	Y	Y	Y	Y	N	Y	Y	Y	Y	N	Y
	C3 regional resources	Y	Y	Y	Y	Y	Y	Y	Y	Y	Y	Y	Y	Y	Y	Y	Y	Y	Y	Y	Y	Y
	C4 regional labour market	Y	Y	Y	Y	Y	Y	Y	Y	Y	Y	Y	Y	Y	Y	Y	Y	Y	Y	Y	Y	Y
	C5 regional economics cycles/regional income generation	Y	Y	Y	Y	Y	Y	Y	Y	Y	Y	Y	Y	Y	Y	Y	Y	Y	Y	Y	Y	Y

of approaches create awareness, sustainability visions and at least different tangible outcomes shown in this chapter. We argue that by using certain methods and tools even more awareness can be raised, social learning strengthened and even more ambitious sustainability strategies adopted and implemented, as can be shown on the provincial approach of Upper Austria that provides 47 sustainability objectives and a detailed indicator system. In addition, the system is flexible as municipalities can choose criteria according to their preferences and local conditions. Therefore, content integration according to the standards laid out earlier cannot be guaranteed which would also call for more science-based approaches – which might conflict with the voluntary nature of process participation.

The second example is the ELAS-calculator (Stoeglehner *et al.*, 2011). The ELAS-calculator is a freely available online-tool (www.elas-calculator.eu) to evaluate settlement projects concerning energy supply and assess environmental and socio-economic effects on the pre-assessment level according to the indicator pyramid. The ELAS-calculator can be used for comprehensive spatial planning, and for LA21 or sectoral energy planning and is especially designed for modelling highly complex planning tasks. By entering a few basic data about a residential settlement (like information about the site, distances to local and regional centres, kind of houses, amount and age structure of inhabitants, energy standards, energy supply and others) the energy demand for room heating and hot water production, electricity and mobility of the inhabitants can be calculated using a long-term perspective (which is critical in achieving sustainable development – see Chapter 4) considering construction and operation of buildings and public infrastructure as well as mobility including embodied energy.

In this way a very complex system of relations between spatial development strategies, building standards, energy supply and certain aspects of energy relevant behaviour of inhabitants can be transferred into a single measure, the energy demand in kWh. The energy demand and supply is used to calculate the Sustainable Process Index of the settlement and the related CO_2 emissions (environmental dimension), regional economic turnovers (regional-economic dimension) and associated employment effects (social dimension) of the settlement. The ELAS-calculator can be used for analysis of existing settlements and has a planning mode for changes in existing, and assessment of new, settlement projects. As changes of the settlement projects can be instantly made visible by changes of the parameters for calculation (e.g. changes in energy supply, effects of thermal insulation, construction of new buildings, increase/decline of inhabitants, change of living space etc.) it serves as an integrated planning and assessment tool. The calculator is available in German and English. As the regional-economic analysis is tailor-made to the Austrian economic situation it cannot be applied to other countries. Therefore it is only provided if sites in Austria are chosen, for other countries only SPI and CO_2 emissions can be calculated.

The ELAS-calculator provides an example of a tool that assists the planning and assessment process as planning alternatives can be generated, instantly assessed and the effects of parameter changes can instantly be made visible even though the systems under evaluation are highly complex. In this way the method is even feasible for deliberative debate as 'playing' with proposals does not require desktop assessment but can be carried out in planning meetings as soon as a computer, a means of projecting the computer output and an internet connection are available. Therefore, the learning processes defined above are facilitated through reflections of values, (proposed) actions and consequences on one or all pillars of sustainable development. In this way both instrumental and conceptual learning are supported. This can be argued with the enormous capacity overshoots current developments cause in all pillars of sustainability which make fundamental changes of visions and values for societal and economic functioning inevitable (see, for example, Kanatschnig and Weber, 1998; Meadows *et al.*, 1992; Jänicke and Jörgens, 2004).

Summing up, from these examples the following methodological challenges for delivering content and process integrated sustainability assessment can be derived. The first example of LA21 shows that already relatively simple lists of criteria give incentives for local communities to contribute to sustainable development. The simplicity of the assessment does, on the one hand, provide planning guidance for bottom-up processes but does not, on the other hand, provide for a structured assessment of certain proposed actions on other sustainability issues. Further methodological development is necessary to add this notion to the planning and assessment process taking the peculiarities of the planning and assessment process into account. These are:

1 highly complex systems
2 subject to a multitude of potential actions addressed to and affecting all pillars of sustainability
3 highly communicative planning processes involving a significant amount of voluntarily engaged non-experts and limited professional workforce.

The necessity to support conceptual learning is, therefore, very clear.

The second example, the ELAS-calculator, provides a practical example to help these deliberative processes. In our understanding of sustainability it provides a best practice example as delivering general indicators for the assessment supports the main features of content integration, and learning processes can be facilitated, as changes in planning proposals can be made instantly visible. Therefore, we suggest that many of these methods, which Stoeglehner (2010b) refers to as 'strategic planning and assessment methods', should be developed in order to facilitate different kinds of comprehensive and sectoral planning and assessment processes in order to approach sustainable development.

Conclusions

Why should sustainability assessment be applied to planning, especially when we consider that many planning schemes strive for sustainability anyhow? And why should sustainability assessment be integrated? The answers to these questions at the beginning can be summarized from the line of argumentation of this chapter: because integrated planning and sustainability assessment processes have the potential to considerably increase the sustainability performance of planning outcomes, depending on the type of plan under assessment. We propose a vision for integrated planning and sustainability assessment processes that try to bridge controversial discourses around sustainability assessment, especially the issue of trade-offs between environmental issues and socio-economic considerations. We are aware that this vision is optimistic and the gap between this vision and assessment reality might, depending on the context, be considerable, but we also believe that this vision is practicable. Whether sustainability assessment will finally become a valuable instrument in planning will depend on the effectiveness of single sustainability assessments to influence the planning outcomes towards sustainable development and to organize participatory debate, if not guaranteed by the planning process itself. Therefore, the methods for assessment and process design will determine whether the sustainability assessment is perceived as being useful, and the success of the sustainability assessment concept.

In order to choose and develop methods and tools for sustainability assessment the decision-making context has to be thoroughly considered, especially the knowledge base, values and interests inherent in the stakeholder groups involved. There are huge differences between sustainability assessments driven by large projects as presented in previous chapters and comprehensive or sectoral planning processes on different levels of government. In order to achieve sustainable development, methods and tools have to be applied that allow for joint planning and assessment and for social learning on the level of facts and the level of values in learning loops between values/visions, actions and consequences. Therefore, sustainability assessment methods and tools have to be put forward to jointly elaborate on the planning, the learning and the assessment process in the development and implementation of planning processes. We hope this chapter gives an incentive for sustainability assessment academics and professionals to engage themselves in the provision of such sustainability assessment methods and tools because they are a prerequisite for planning processes and the accompanying sustainability assessment to lead to actual sustainable development.

Acknowledgements

The analysis of the Austrian LA21 strategies was carried out within the research in the 'Vital Landscapes' project (Project Nr. 2CE164P3, www.vital-landscapes.eu) that is implemented through the CENTRAL EUROPE programme (www.central2013.eu) and co-financed by the European Rural Development Fund.

The above presented ELAS calculator was developed within the framework of the project 'ELAS – Energetic Long-term Analysis of Settlement Structures' that was funded by the Austrian Climate and Energy Fund (www.klimafonds.gv.at) within the program line 'Neue Energien 2020' (Project Nr. 818915) and co-financed by the Provincial governments of Upper and Lower Austria as well as the municipality of Freistadt.

References

Argyris C (1993) *Knowledge for Action: A Guide to Overcoming Barriers to Institutional Change* (San Francisco: Jossey-Bass).

Birkmann J (2000) 'Nachhaltige Raumentwicklung im dreidimensionalen Nebel', *UVP-Report,* 14(3), 164–167.

Brown AL and Thérivel R (2000) 'Principles to guide the development of SEA methodology', *Impact Assessment and Project Appraisal,* 18(3), 183–190.

Dalkmann H (2005) 'Die Integration der Strategischen Umweltprüfung in Entscheidungsprozesse', *UVP-Report,* 19(1), 31–34.

Elling B (2009) 'Rationality and effectiveness: does EIA/SEA treat them as synonyms?', *Impact Assessment and Project Appraisal,* 27(2), 121–132.

European Parliament and the Council of the European Union (2001) 'Directive 2001/42/EC of the European Parliament and of the Council of 27 June 2001 on the assessment of the effects of certain plans and programmes on the environment', *Official Journal of the European Communities,* L197, pp.30–37.

ExpertInnengruppe LA21 (2010) 'LA21-Basisqualitäten 3.0, Prozessorientierte, partizipative und inhaltliche Basisqualitäten für Lokale Agenda 21-Prozesse in Österreich ab 2009'. Positionspapier der ExpertInnengruppe Dezentrale Nachhaltigkeitsstrategien – Lokale Agenda 21.

Fischer TB, Wood C, Jones C (2002) 'Improving the practice of policy, plan and programme environmental assessment', *Environment and Planning B,* 29(2), 159–172.

Fürst D and Scholles F (2001) *Handbuch Theorien + Methoden der Raum-und Umweltplanung* (Dortmund: Dortmunder Vertrieb für Bau- und Planungsliteratur).

Grühn D (2004) *Zur Validität von Bewertungsmethoden in der Landschafts- und Umweltplanung* (Berlin: Habilitationsschrift, Technische Universität Berlin).

Hacking T and Guthrie P (2008) 'A framework for clarifying the meaning of Triple Bottom-Line, Integrated, and Sustainability Assessment', *Environmental Impact Assessment Review,* 28, 73–89.

Healey P (1992) 'Planning through debate. The communicative turn in planning theory', *Town Planning Review,* 63, 143–162.

Innes J and Booher DE (2000) *Collaborative Dialogue as a Policy Making Strategy. Institute of Urban and Rural Development, University of California, Berkeley – Working Paper Series.* [Online] Available at: <http://repositories.cdlib.org/iurd/wps/WP-2000–05> (accessed 31 October 2011).

Jänicke M and Jörgens H (2004) 'Neue Steuerungskonzepte in der Umweltpolitik', *ZfU* 3/2004, 297–348.

Kanatschnig D and Weber G (1998) *Nachhaltige Raumentwicklung in Österreich.* (Wien: Österreichisches Institut für Nachhaltige Entwicklung).

Meadows DH, Meadows DL, Randers J (1992) *Die neuen Grenzen des Wachstums. Die Lage der Menschheit: Bedrohung und Zukunftschancen* (Stuttgart: Deutsche Verlags Anstalt).

Morrison-Saunders A and Fischer T (2006) 'What is wrong with EIA and SEA anyway? A sceptic's perspective on sustainability assessment', *Journal of Environmental Assessment Policy and Management,* 8(1), 1–21.

Mueller S (2004) 'Internationale Einflüsse auf die Planungstheoriedebatte in Deutschland nach 1945 oder die Perspektiven der Planungsdemokratie'. In: Altrock U, Guenther S, Hunning S, Peter D (eds) *Perspektiven der Planungstheorie* (Berlin: Leue) 123–140.

Narodoslawsky M and Stoeglehner G (2010) 'Planning for local and regional energy strategies with the ecological footprint', *Journal of Environmental Policy & Planning,* 12(4), 363–379.

Partidário MR (1996) 'Strategic environmental assessment: key issues emerging from recent practice', *Environmental Impact Assessment Review,* 16, 31–55.

Partidário MR, Paddon M, Eggenberger M, Chau DM, Duyen NV (2008) 'Linking strategic environmental assessment (SEA) and city development strategy in Vietnam', *Impact Assessment and Project Appraisal,* 26(3), 219–227.

Pope J and Grace W (2006) 'Sustainability assessment in context: issues of process, policy and governance', *Journal of Environmental Assessment Policy and Management,* 8(3), 373–398.

Pope J, Morrison-Saunders A, Annandale D (2005) 'Applying sustainability assessment models', *Impact Assessment and Project Appraisal,* 23(4), 293–302.

Retief F (2007) 'Effectiveness of Strategic Environmental Assessment (SEA) in South Africa', *Journal of Environmental Assessment Policy and Management,* 9(1), 83–101.

Richardson T (2005) 'Environmental assessment and planning theory: four short stories about power, multiple rationality and ethics', *Environmental Impact Assessment Review,* 25, 341–365.

Runhaar H and Driessen P (2007) 'What makes strategic environmental assessment successful environmental assessment? The role of context in the contribution of SEA to decision making', *Impact Assessment and Project Appraisal,* 25(1), 2–14.

Scharpf F (2000) *Interaktionsformen. Akteurszentrierter Institutionalismus in der Politikforschung* (Opladen: Leske + Budrich).

Stoeglehner G (2004) 'Integrating strategic environmental assessment into community development plans – a case study from Austria', *Journal of European Environmental Policy,* 14, 58–72.

Stoeglehner G (2009) *Von der Umweltprüfung zur Umweltplanung – Reflexionen anhand der Raumplanung in Österreich* (Vienna: Habilitationsschrift, Universität für Bodenkultur).

Stoeglehner G (2010a) 'Enhancing SEA effectiveness: lessons learnt from Austrian experiences in spatial planning', *Impact Assessment and Project Appraisal,* 28(3), 217–231.

Stoeglehner G (2010b) 'SUP und Strategie – eine Reflexion im Lichte strategischer Umweltprobleme', *UVP-Report*, 23(5), 262–266.

Stoeglehner G and Narodoslawsky M (2008) 'Implementing ecological footprinting in decision-making processes', *Land Use Policy*, 25, 421–431.

Stoeglehner G, Brown AL, Kørnøv LB (2009) 'SEA and planning: "ownership" of strategic environmental assessment by the planners is the key to its effectiveness', *Impact Assessment and Project Appraisal*, 27(2), 111–120.

Stoeglehner G, Mitter H, Weiss M, Neugebauer G, Narodoslawsky M, Niemetz N, Kettl K-H, Baaske W, Lancaster B (2011) ELAS – Energetische Langzeitanalyse von Siedlungsstrukturen. Projekt gefördert aus Mitteln des Klima- und Energiefonds, der Länder Oberösterreich und Niederösterreich sowie der Stadt Freistadt.

Thérivel R and Minas P (2002) 'Ensuring effective sustainability appraisal', *Impact Assessment and Project Appraisal*, 20(2), 81–91.

Thérivel R, Christian G, Craig C, Grinham R, Mackins D, Smith J, Sneller T, Turner R, Walker D, Yamane M (2009) 'Sustainability-focused impact assessment: English experiences', *Impact Assessment and Project Appraisal*, 27(2), 155–168.

United Nations (1992) *Agenda 21: Earth Summit, The United Nations Programme of Action from Rio* (United Nations Publications: Rio de Janeiro). [Online] Available at: <http://www.un.org/esa/dsd/agenda21/index.shtml> (accessed 31 October 2011).

Weiland U and Wohlleber-Feller S (2007) *Einführung in die Raum- und Umweltplanung* (Paderborn: UTB).

Willis MR and Keller AA (2007) 'A framework for assessing the impact of land use policy on community exposure to air toxics', *Journal of Environmental Management*, 83(2), 213–227.

17 Conclusions

Alan Bond, University of East Anglia
Angus Morrison-Saunders, Murdoch University and North West University
Richard Howitt, Macquarie University

Introduction

Sustainability assessment is a vibrant and engaging activity that seeks direct actions and decisions towards sustainability. It can be applied to a wide range of activities and involves a broad spectrum of stakeholders. Sustainability assessment is challenging with respect to theory, process and practice. But equally it is a rewarding and essential undertaking to initiate the necessary shift towards positive sustainability behaviours.

Like the fields of impact assessment from which it builds, sustainability assessment is intended to change things. Fundamentally, it is a way of evaluating decisions, projects and processes that allows decision makers in governments, companies and communities to secure outcomes and opportunities that shift a range of human actions away from paths of demonstrably unsustainable – and therefore inappropriate – relationships with and impacts on natural and social systems. Given the dynamics of both natural and human systems, of course, sustainability assessment cannot give simple guarantees that a particular decision, project or proposal will be 'sustainable'. Not only does the context of pluralism mean that the content of concepts such as 'sustainability' will always be socially, culturally and politically contested, but also the goal of sustainability itself is a moving target. Sustainability is not a state that once achieved could be checked off as 'done' and left behind. Sustainability assessment offers tools and processes to monitor the extent to which human decisions are producing or contributing to unsustainability, and the efficacy of efforts to be sustainable at various spatial and temporal scales specified as relevant in a particular context.

In this chapter we reflect on the previous 16 chapters of the book (and the preface) and consider what is offered in terms of helping to design a better sustainability assessment process. The book is presented in four distinct parts:

- the first part explained why sustainability assessment is needed and what theoretical frameworks might be considered when evaluating practice
- the second part examined pluralism, arguing that pluralism was a critical issue in conducting sustainability assessment

- the third part examined practice but, in doing so, first needed to establish a framework against which practice could be judged. The framework drew on Parts 1 and 2 of the book
- the final part addressed key challenges and problems facing sustainability assessment and discussed what might be needed to secure improved practice in this important field of impact assessment and policy making.

To frame the remainder of this chapter, we raise a number of questions intended to guide critical reflection on the pluralism of sustainability assessment and the contributions made in this book:

1 Did the focus on pluralism throughout this book adequately represent the key issues that need to be taken into account when evaluating sustainability assessment practice?
2 How robust is the evaluation framework developed in Chapter 8 for judging (and improving) sustainability assessment procedures and practices worldwide?
3 To what extent are the countries evaluated in Part 3 of the book sufficiently representative of practice (this is important because it is problems identified with this sample of four countries that has led to the recommendations for improvement in Part 4)?
4 What key challenges need to be considered when designing or implementing sustainability assessments?
5 Are the recommendations made in Part 4 realistic and helpful for advancing the theory, processes and practice of sustainability assessment?

The rest of this chapter is structured around answering these questions. One issue we did consider is the fact that, despite being a book on sustainability assessment, we do not set out a particular process to follow. By way of contrast, textbooks on Environmental Impact Assessment (EIA), Social Impact Assessment (SIA) or Strategic Environmental Assessment (SEA) commonly set out a specific process that if followed should produce an adequate assessment outcome (see, for example, Wathern, 1988, p.18; Canter, 1996, p.38; Becker, 1997, pp.7–8; Becker and Vanclay, 2003, pp.8–9; Glasson *et al.*, 2005, p.4; Fischer, 2007, p.4; Hanna, 2009, pp.9–13). Nowhere does our book prescribe for practitioners how a sustainability assessment should be executed, nor what stages should be included. This was a deliberate decision rather than an accidental oversight or omission. We debated this matter intently when proposing and designing the book, and judged that it would be inappropriate and misleading to offer up a 'model-based' approach to sustainability assessment. There is no checklist of questions, methods and steps that will work in all circumstances because the specific context of any assessment exercise matters. Our decision was taken because pluralism is central to good sustainability assessment. In our view, this extends not only to views and beliefs about sustainability, but also to governance processes, and sustainability assessment practice, including whether it should be legally mandated or voluntary. Different people rightly will have different views on how sustainability assessment should be realised. Ultimately, too, the specific detail of an approach, we believe, is context-specific and our judgement is that the research drawn into the discussion in this book convincingly demonstrates that this understanding is correct.

The focus on pluralism

There is a clear focus on pluralism throughout this book. This was driven by our understanding, developed on the basis of our experience in doing impact and sustainability assessment research in a variety of settings over the past 20 years, that people matter. Considering pluralism to be central to successful decision making acknowledges that there are different views and beliefs about what decisions should be, and what constitutes a successful decision-making process. Such differences, of course, are not just because different people might be unevenly affected by a particular project. Cultural differences, for example, constitute what is an appropriate (and sustainable) relationship between human society and the natural world; they constitute the ethical dimensions of intergenerational responsibility differently; they develop profoundly different understandings of risk, responsibility and value. And sustainability assessment must address that pluralism as the foundation for making judgements and decisions.

This understanding is in step with the increasing emphasis on public participation and community engagement that pervades public policy making around the globe, and can be traced back to seminal publications identifying different levels of participation (see, in particular, Arnstein, 1969) which improved understanding of the extent to which assessment and decision-making practices were marginalising citizens. There was an increasing realisation that when decision making was left as the dominion of the political, economic and professional elites who were typically insensitive to the rights and hopes of populations directly affected by their decisions, the decision processes themselves were producing bad decisions and negative impacts in terms of social, environmental and sustainability outcomes (Vanclay, 2004). This has led to much greater emphasis on earlier, and better public engagement, particularly within decision-making tools such as EIA, SIA and SEA.

Our focus on pluralism is complemented by our adoption of a constructivist theory of knowledge that holds that people generate knowledge and meaning from their shared (and contested) experiences and beliefs. While other theories of knowledge exist, in making judgements and decisions based on sustainability assessments, it is in working through the issues that arise from the shared and contested fields of experience and beliefs that the hard work needs to be done. That capitalist economies and large-scale economic decisions by governments and international agencies tend to be driven by materialism (i.e. values and understandings about the generation and distribution of wealth) is part of the sustainability conundrum. Indeed, diversity of ways of knowing is precisely part of the pluralism the book addresses.

A robust evaluation framework

The evaluation framework developed in Chapter 8 is a framework for thinking about effectiveness for assessment processes. Whilst it has been developed based on a review of existing literature, it is very much a research-informed judgement at a point in time. We accept that others have different opinions and reach different conclusions, and that our own knowledge, understanding and conclusions are likely to change over time as we learn more about sustainability assessment and as the challenge of sustainability changes over time. The questions we have asked have collated a significant number of ideas for each category of effectiveness, and therefore allow some level of interpretation by the authors of the practice chapters (Chapters 9–12). Sustainability assessment practice is so different in the jurisdictions covered in this book that we considered providing such a framework to be the only approach that would allow any degree of cross comparison.

The application of effectiveness criteria also suggests some anticipated ideal outcome, and therefore is judgemental when applied. Even when trying to take a constructivist approach, this means imposing some level of framing on those who apply the framework. The most obvious example in the framework is the consideration of normative effectiveness to encompass the imperatives of sustainability set out by Gibson in Chapter 1. The notion of 'norms' suggest thinking which is acknowledged or accepted in some way, but at the same time, accepting the importance of pluralism suggests that these norms vary between individuals and organisations. Our framework imposes our norms (as editors) – which is, of course, at odds with an argument that sustainability goals should be broadly agreed by stakeholders at the outset. We acknowledge this apparent paradox and weakness. To be able to compare and appraise practice at all does require some means of evaluation, and when undertaken in a consistent way, this does need an approach like the one we have taken.

Moving forward in sustainability assessment certainly requires practitioners to make judgements about the effectiveness of specific sustainability assessment approaches, reports and recommendations. The framework developed here offers a benchmark from which to judge and to act. The framework itself will not resolve divergent theories of knowledge, but it identifies important questions and frames values and principles in a way that facilitates debate across differences. In this way, the framework is intended to facilitate communication, rather than to present a once-and-forever solution.

So, is the framework robust? We would argue it is a sound means for comparison, but would not make claims that it allows us to draw definitive conclusions as to the 'effectiveness' of any process to which it applies. It is a useful device to identify important aspects which should be considered in process design; it is a thought-starter for deliberation by sustainability assessment stakeholders.

Representing sustainability assessment practice

The argument was made early in the book that the four jurisdictions discussed in Part 3 were included because they represented examples of sustainability assessment practice at the time of writing. However, when reflecting on this argument and based on the practice the chapters describe, our initial proposition seems rather tenuous. In England, there is a legal requirement for sustainability appraisal, and in Western Australia there have been voluntary processes conducted under the name of sustainability assessment. In Canada, practice is variable between the different territories, which makes it complicated to generalise about 'Canadian practice'. In South Africa, the fact that sustainability assessment is conducted is based on an interpretation of the existing laws in the country to the extent that they encompass sustainability principles. Similar claims can probably be made for other countries – after all, any of the impact assessment tools in use worldwide can arguably be considered to be directed towards sustainable development at some level. The truth is, therefore, that some of the examples of practice have been identified because previous academic research has framed practice in those localities as sustainability assessment, whereas in other countries where impact assessment practice may also be described in similar terms, equivalent claims have yet to be publicly declared.

So, the comparison in Chapters 9–12 is useful, but should not be taken to be representative of the breadth and depth of sustainability assessment undertakings worldwide. We look forward to future discussion of additional examples of good and improving practice in other jurisdictions (including international arenas), and would encourage readers to use the framework offered here to develop their own systematic approach to reviewing the effectiveness of impact assessment and related processes as a basis for securing more sustainable outcomes in public decision making.

What key challenges face sustainability assessment?

In the final part of the book, we have aimed to articulate the most significant challenges facing sustainability assessment at this point in its development, and to frame discussion of those challenges in the light of the demands of pluralism and practice. The structure (and hence content to a large extent) of Part 4 of the book was planned before any of the other parts were written. This did not mean that we had already reached our own conclusions and recommendations on sustainability assessment in isolation from the research and thinking presented in the book. Simply, such an approach was necessitated by the contractual arrangements involved in preparing a book for publication, where some certainty is needed about authors, content and length. So in advance, we had decided that the key areas that needed attention to improve practice would be engagement and learning, better process and integration into decision-making process (so the sustainability assessment was influential in some way).

The attention to engagement and learning is, on reflection, a choice we are very happy with. These chapters acknowledge the importance of accommodating pluralism, and also the imperfections in any assessment process, which are only likely to get better if practitioners actively acknowledge the need to learn how to do them better, and to correct mistakes that will inevitably be made along the way. The chapter on process allowed us to identify the interlinkages between the various effectiveness criteria, and this, to an extent, justified the focus on engagement and learning. It also indicated that good process design is crucial to achieving good process. The chapter on integration illustrates how context-dependent sustainability assessment is. Once the principles we developed in the book are taken forward to consider exactly how a process might integrate with decision making, then many decisions will already have been made, in particular, about the sustainability framing that will be adopted.

We feel our choice in framing the key challenges has been largely vindicated, but we also recognise that rapid development in fields as diverse as economic crisis, climate risk and natural hazards means that new challenges to the practice and application of sustainability assessment will emerge rapidly in the future. We therefore encourage readers to reflect on what they judge might be missing from Part 4 in terms appropriate to their own circumstances in time and space, their own cultural, social, political and environmental setting. We would be delighted to respond to debate on their conclusions, but at this point we conclude that we would not have included any other chapters based on the information gained on practice. In the future, as we learn more, this might change.

Are the recommendations made in Part 4 realistic?

There is a danger, when making recommendations, to expect so much to be achieved through the process of sustainability assessment, that transactive effectiveness will not be achieved. The reality of the modern world is that assessment costs money and takes time, and there will never be enough money or enough time to conduct the level of assessment that might be considered ideal. It is also true that levels of uncertainty in economic, environmental and political realms is going to mean that any specific recommendations about what might be 'ideal' in any given setting will be both hard to pin down, and contested by multiple stakeholders.

Our recommendations no doubt have financial implications in the short term for proponents and practitioners, relative to traditional forms of impact assessment. Good engagement is not free, but the key principles are to be transparent and inclusive – and these principles don't cost money. We believe that the systemic benefits from improved sustainability assessment practices (for both natural and human systems) are substantial, demonstrable and desirable – and that approaches can be found which reduce the financial burdens that might arise for particular stakeholders (e.g. proponents or governments). It is, in any case, a reasonable societal expectation that a project, plan or policy, which is ultimately materialistic (as they often are), should be required and should expect to pay for an appropriate level of engagement, and should be held accountable for its implications for future consequences. Our recommendations relating to learning rely heavily on reflection, and in practitioners, proponents, communities and researchers being prepared to take the appropriate actions arising from their reflections. That reflection will sometimes translate into doing things differently in the next sustainability assessment, but we have also included an expectation that the reflection can also lead to remediation of unforeseen impacts, through some forms of adaptive management, and even to ongoing assessment tasks or processes to some extent. In recognising the significance for sustainability of pluralism, the urgent need for systems that are adaptive to new insights across cultural diversity has been recognised in this book, to counter political economic and legislative systems previously predicated on denying and devaluing the importance of ecological and cultural diversity. The costs of such adaptation might be unforeseen, but our argument is that it is not unreasonable – indeed, in reorienting societies towards sustainability, they may be simple necessities. A parallel might be considered from the consumer society in which we live. If we purchase a (reasonably expensive) good, we expect some form of product support and a guarantee. If something goes wrong, then the guarantee needs to be honoured, and the manufacturer has to embed the price of honouring the guarantee into the price charged for the good. The same principle can surely be applied when considering, for example, resource extraction, with impact and sustainability assessment providing the information on which warranties are based and executed.

For us as editors, the bottom line is that if sustainability assessment is going to lead to legitimate and successful decisions, the sustainability assessment needs to be properly resourced. But resources alone are insufficient to secure and support the changes needed. Nor is better practice in sustainability assessment. This book has demonstrated just how important a shift in societal and institutional understanding of sustainability principles and the development of robust, effective and affordable sustainability assessment processes as a normal part of normal decision making in complex pluralist human systems are. It has offered readers an opportunity to consider basic principles, important experiences, effective methods and significant challenges to integrate into emerging practice. We are grateful to our contributors for their work in shaping our thinking and offering discursive openings for readers to pursue their own debates and interests. We look forward to seeing the results of such discourses, and having to respond to them, in the future.

References

Arnstein SR (1969) 'A ladder of citizen participation', *Journal of the American Institute of Planners*, 35(4) 216–244.

Becker H (1997) *Social Impact Assessment: Method and Experience in Europe, North America and the Developing World* (London: UCL Press).

Becker H and Vanclay F (eds) (2003) *The International Handbook of Social Impact Assessment: Conceptual and Methodological Advances*. (Cheltenham: Edward Elgar).

Canter LW (1996) *Environmental Impact Assessment* (New York: McGraw-Hill).

Fischer TB (2007) *Theory & Practice of Strategic Environmental Assessment* (London: Earthscan).

Glasson J, Thérivel R, Chadwick A (2005) *Introduction to Environmental Impact Assessment* (Abingdon: Routledge).

Hanna KS (ed.) (2009) *Environmental Impact Assessment: Practice and Participation*. (Ontario: Oxford University Press).

Vanclay F (2004) 'The triple bottom line and impact assessment: how do TBL, EIA, SIA, SEA and EMS relate to each other?', *Journal of Environmental Assessment Policy and Management*, 6(3) 265–288.

Wathern P (1988) *Environmental Impact Assessment: Theory and Practice* (London: Routledge).

Index